ISBN 978-1-332-75930-9
PIBN 10433910

THEORY OF STEAM ENGINE.

Translated from the fourth edition of Weisbach's Mechanics, By A. J. DuBois. Containing notes giving practical examples of Stationary, Marine, and Locomotive Engines, showing American practice, by R. H. Buel. Numerous illustrations.
8vo, cloth, $5 00

THE STEAM ENGINE CATECHISM.

A Series of direct practical answers to direct practical questions, mainly intended for young engineers, and for examination questions. By Robert Grimshaw. Fifth edition, enlarged and improved. 1887............ ....18mo, cloth, 1 00

"Not only young Engineers, but all who desire rudimental and practical instruction in the science of Steam Engineering will find profit in reading the 'Steam Engine Catechism' by Robt. Grimshaw."—*Mechanical News.*

STEAM ENGINE CATECHISM. Part II.

Containing answers to further practical inquiries received since the issue of the first volume. Second edition, enlarged.
18mo, cloth, 1 00

"It deserves a place in every young engineer's book-case."—*Engineering*, London.

STATIONARY STEAM ENGINES.

Especially adapted to Electric Lighting Purposes. Treating of the Development of Steam Engines—the principles of Construction and Economy, with description of Moderate Speed and High Speed Engines. By Prof. R. H. Thurston.
12mo, cloth, 1 50

"This work must prove to be of great interest to both manufacturers and users of steam engines."—*Builder and Wood Worker.*

INDICATOR PRACTICE AND STEAM ENGINE ECONOMY.

With Plain Directions for Attaching the Indicator, Taking Diagrams, Computing the Horse-power, Drawing the Theoretical Curve, Calculating Steam Consumption, Determining Economy, Locating Derangement of Valves, and making all desired deductions; also, Tables required in making the necessary computations, and an Outline of Current Practice in Testing Steam Engines and Boilers. By Frank F. Hemenway, Associate Editor "American Machinist," Member American Society Mechanical Engineers, etc...............12mo, cloth, 2 00

"The most interesting book on Steam Engineering that we have ever read."—*National Car and Locomotive Builder.*

"Should be in the hands of every engineer."—*Engineering News.*

"Must be a boon to the every-day engineer who is seeking information."—*Sanitary Engineer.*

TWENTY YEARS WITH THE INDICATOR.

By Thos. Pray, Jr., C.E., and M.E........ 2 vols., 8vo, cloth, 3 00
Volume I., 8vo, cloth, 1 50
Volume II., 8vo, cloth, 2 00

This work is by a practical engineer of twenty-four years' experience in adjusting all kinds of engines, from the smallest portable to new and largest locomotives and ocean steamships up to 1885.

MARINE ENGINES AND DREDGING MACHINERY.

Showing the latest and best English and American Practice. By Wm. H. Maw. Illustrated by over 150 fine steel plates, (mostly two-page illustrations), of the engines of the leading screw steamships of England and other nations, and 295 fine wood engravings in text...... ....folio, half morocco, 18 00

"A superb volume."

TABLES, WITH EXPLANATIONS, RELATING to the NON-CONDENSING STATIONARY STEAM ENGINE, and of HIGH-PRESSURE STEAM BOILERS.

By W. P. Trowbridge. Plates...........4to, paper boards, 2 50

TABLES

OF THE PROPERTIES OF

SATURATED STEAM

AND OTHER VAPORS.

BY

CECIL H. PEABODY, B.S.,

ASSISTANT PROFESSOR OF STEAM ENGINEERING IN THE MASSACHUSETTS INSTITUTE OF TECHNOLOGY.

NEW YORK:
JOHN WILEY AND SONS,
1888.

RAND AVERY COMPANY,
ELECTROTYPERS AND PRINTERS,
BOSTON.

SATURATED STEAM, AND OTHER VAPORS.

A COMPARISON of the several tables of the properties of saturated steam, expressed in English units, reveals discrepancies of considerable magnitude; and investigation shows that, while all are in some manner founded on the experiments of Regnault, various methods of calculation have been used, and in some cases other experimental data have been employed. A rêview of the whole subject, in connection with the preparation of notes on thermodynamics for the use of the students of the Massachusetts Institute of Technology, made it seem important to calculate a set of tables, to accompany those notes, founded on the best and most recent data.

In presenting the tables for general use, it appears proper to state in full the data and the methods of calculation employed, so that each one may see the degree of accuracy and correctness of the tables, and the reliance to be placed on them.

Tables of the properties of other vapors have been added, which will be discussed hereafter.

Pressure of Saturated Steam. — As a conclusion from all the experiments on the tension of saturated steam, Regnault gives, in the *Mémoires de l'Institut de France, etc., Tome XXI.*, the following data: —

TEMPERATURE C.	PRESSURE MM. OF MERCURY.
−32	0.32
−16	1.29
0	4.60
25	23.55
50	91.98
75	288.50
100	760.00
130	2030.0
160	4651.6
190	9426.
220	17390.
−20	0.91
+40	54.91

From these data he calculated, by the aid of seven-place logarithms, the following formulæ, which give the pressure in millimetres of mercury for any temperature in degrees Centigrade: —

A. For steam from $-32°$ to $0°$ C.

$$p = a + b\alpha^n.$$
$$a = -0.08038.$$
$$\log b = 9.6024724 - 10.$$
$$\log \alpha = 0.033398.$$
$$n = 32° - t.$$

B. For steam from $0°$ to $100°$ C.

$$\log p = a - b\alpha^n + c\beta^n$$
$$a = 4.7384380.$$
$$\log b = 0.6116485.$$
$$\log c = 8.1340339 - 10.$$
$$\log \alpha = 9.9967449 - 10.$$
$$\log \beta = 0.006865036.$$
$$n = t.$$

C. For steam from $100°$ to $220°$ C.

$$\log p = a - b\alpha^n + c\beta^n$$
$$a = 5.4583895.$$
$$\log b = 0.4121470.$$
$$\log c = 7.7448901 - 10.$$
$$\log \alpha = 9.997412127 - 10.$$
$$\log \beta = 0.007590697.$$
$$n = t - 100.$$

D. For steam from $-20°$ to $220°$ C.

$$\log p = a - b\alpha^n - c\beta^n.$$
$$a = 6.2640348.$$
$$\log b = 0.1397743.$$
$$\log c = 0.6924351.$$
$$\log \alpha = 9.994049292 - 10.$$
$$\log \beta = 9.998343862 - 10.$$
$$n = t + 20.$$

By aid of the formulæ A and B, Regnault calculated and recorded tables of the pressures of saturated steam for temperatures from $-32°$ to $100°$ C. The formula D was calculated from the data given above for the temperatures $-20°$, $+40°$, $100°$, $160°$, and $220°$ C., and was intended to represent the whole range of experiments. By this formula, instead of formula C, he calculated the pressures set down in his tables for temperatures from $100°$ C. to $220°$ C.

Wishing to obtain greater accuracy for meteorological work, Moritz recalculated Equation B, using ten-place logarithms, and obtained constants that differ but little from those that will be given later. Some of the more recent tables in the French system were calculated by his equations.

Equations for the Pressure of Steam at Paris. — In view of the preceding statements, it appeared desirable to re-calculate the constants for Equations *B* and *C*, with a degree of accuracy that should exclude any doubt as to the reliability of the results. Accordingly, the logarithms required were taken from Vega's ten-place table, and then the remainder of the calculations were carried on with natural numbers, checking by independent methods, with the following results: —

B. For steam from 0° to 100° C.

$$\log p = a - b\alpha^n + c\beta^n.$$
$$a = 4.7393622142.$$
$$\log b = 0.6117400190.$$
$$\log c = 8.1320378383 - 10.$$
$$\log \alpha = 9.996725532820 - 10.$$
$$\log \beta = 0.006864675924.$$
$$n = t.$$

C. For steam from 100° to 220° C.

$$\log p = a - b\alpha^n + c\beta^n.$$
$$a = 5.4574301234.$$
$$\log b = 0.4119787931.$$
$$\log c = 7.7417476470 - 10.$$
$$\log \alpha = 9.99741106346 - 10.$$
$$\log \beta = 0.007642489113.$$
$$n = t - 100.$$

To show the degree of accuracy attained, the following tables are given: —

Equation *B*.

t.	*p*.	LOG *p* FROM TABLE OF LOGARITHMS.	LOG *p* CALCULATED BY EQUATION.
0	4.60	0.6627578317	. . . . . .
25	23.55	1.3719909115	1.37199097
50	91.98	1.9636934052	1.96369346
75	288.50	2.4601458175	2.46014587
100	760	2.8808135923	2.88081365

Equation *C*.

t.	*p*.	LOG *p* FROM TABLE OF LOGARITHMS.	LOG *p* CALCULATED BY EQUATION.
100	760.00	2.8808135923	. . . .
130	2030.0	3.3074960379	3.307496036
160	4651.6	3.6676023618	3.667602359
190	9426	3.9743274354	3.974327428
220	17390	4.2402995820	4.240299575

The results from Equation C are quite satisfactory; for the errors come in the ninth place of decimals, and one place of decimals is unavoidably lost in the application of the formula. Equation B was calculated after Equation C and the numerical work was not carried to so large a number of decimal places. For the calculation of tables, the constants are carried to seven places of significant figures only; this gives six significant figures in the result, of which five are recorded in the table.

Pressure of Steam at Latitude 45°. — French System. — It is customary to reduce all measurements to the latitude of 45°, and to sea-level. The standard thermometer should then have its boiling and freezing points determined under, or reduced to such conditions. The value of g, the acceleration due to gravity, is, at Paris, latitude 48° 50′ 14″ and 60 metres above sea-level, 9.809218 metres; and at 45°, and at sea-level, it is 9.806056 metres. Consequently, 760 mm. of mercury at 45° gives a pressure equal to that of 759.755 mm. at Paris; and this corresponds to a temperature of 99.°991 C.

In other words, the thermometer which is standard at 45° has each degree 0.99991 of the length of the degree of a thermometer standard at Paris.

To reduce Equation B to 45° latitude, we have

$$\log p = a + \log \frac{980.9218}{980.6056} - b\alpha^{0.99991t} + c\beta^{0.99991t};$$

and for Equation C

$$\log p = a + \log \frac{980.9218}{980.6056} - b\alpha^{(0.99991t - 100)} + c\beta^{(0.99991t - 100)}$$

$$= a + \log \frac{980.9218}{980.6056} - b\alpha^{-0.009}\alpha^{0.99991(t-100)} + c\beta^{-0.009}\beta^{0.99991(t-100)}$$

The resulting equations which were used in calculating Table III are

B. For steam from 0° to 100° C. at 45° latitude.

$$\log p = a_1 - b\alpha_1^n + c\beta_1^n.$$
$$a_1 = 4.739502.$$
$$\log b = 0.6117400.$$
$$\log c = 8.13204 - 10.$$
$$\log \alpha_1 = 9.996725828 - 10.$$
$$\log \beta_1 = 0.0068641.$$
$$n = t.$$

C. For steam from 100° to 220° C. at 45° latitude.

$$\log p = a_1 - b_1\alpha_1^n + c_1\beta_1^n.$$
$$a_1 = 5.457570.$$
$$\log b_1 = 0.4120021.$$
$$\log c_1 = 7.74168 - 10.$$
$$\log \alpha_1 - 9.997411296 - 10.$$
$$\log \beta_1 = 0.0076418.$$
$$n = t - 100.$$

Pressure of Steam at Latitude 45°. — English System. — To reduce the equations for the pressure of steam, so that they will give the pressures in pounds on the square inch for degrees Fahrenheit, there are required the comparison of measures of length, and of weight, the comparison of the scales of the thermometers, and the specific gravity of mercury.

Professor Rogers (*Proceedings of the Am. Acad. of Arts and Sciences, 1882–83*, also *Additional Observations*, etc.) gives for the length of the metre, 39.3702 inches. This differs from the value given by Capt. Clarke (*Proceedings of the Royal Society, vol. xv., 1866*), by an amount that does not affect the values in the tables; his value being 39.370432 inches.

Professor Miller (*Phil. Transactions, cxlvi., 1856*) gives for the weight of one kilogram, 2.20462125 pounds.

Regnault gives, for the weight of one litre of mercury, 13.5959 kilograms.

The degree Fahrenheit is $\frac{5}{9}$ of the length of the degree Centigrade.

Let $$k = \frac{13.5959 \times 2.204621}{\overline{39.3702}^2};$$

then the equations B and C have for the reduction to degrees Fahrenheit, and pounds on the square inch,

$$\log p = a_1 + \log k - b\alpha^{\frac{5}{9}n} + c\beta^{\frac{5}{9}n},$$

$$\log p = a_1 + \log k - b_1\alpha_1^{\frac{5}{9}n} + c_1\ \beta_1^{\frac{5}{9}n}.$$

The resulting equations, which were used in calculating Tables I and II, are: —

B. For steam from 32° to 212° F., in pounds on the square inch.

$$\begin{aligned} \log p &= a_2 - b\alpha_2^n + c\beta_2^n. \\ a_2 &= 3.025908. \\ \log b &= 0.6117400. \\ \log c &= 8.13204 - 10. \\ \log \alpha_2 &= 9.998181015 - 10. \\ \log \beta_2 &= 0.0038134. \\ n &= t - 32. \end{aligned}$$

C. For steam from 212° to 428° F., in pounds on the square inch.

$$\begin{aligned} \log p &= a_2 - b_1\alpha_2^n + c_1\beta_2^n. \\ a_2 &= 3.743976. \\ \log b_1 &= 0.4120021. \\ \log c_1 &= 7.74168 - 10. \\ \log \alpha_2 &= 9.998561831 - 10. \\ \log \beta_2 &= 0.0042454. \\ n &= t - 212. \end{aligned}$$

All of the foregoing equations make the pressure a function of the temperature on the scale of the air-thermometer. It will be assumed that the difference between that scale and the absolute scale may be neglected.

Other Equations for the Pressure of Steam.—Rankine, in his *Steam Engine and other Prime Movers*, gives the following equation:—

$$\log p = A - \frac{B}{T} - \frac{C}{T^2}.$$

For pounds on the square inch, corresponding to degrees Fahrenheit,—

$$A = 6.1007.$$
$$\log B = 3.43642.$$
$$\log C = 5.59873.$$
$$T = t + 461.^\circ 2 \text{ F}.$$

This equation has been largely used for calculating tables on the English system. The following table will give a comparison between the results from this formula and those from Formulæ *B* and *C*.

TEMPERATURE.	PRESSURE.	
	Regnault at 45° latitude.	Rankine.
32	0.0890	0.083
77	0.4555	0.452
122	1.7789	1.78
167	5.579	5.58
212	14.99	14.70
257	33.711	33.71
302	69.27	69.21
347	129.79	129.8
392	225.56	225.9
428	336.26	336.3

Differential Co-efficient $\frac{dp}{dt}$.—As will be seen later, the differential co-efficient $\frac{dp}{dt}$ is used in calculating the volume and density of saturated vapors.

From the general equation of the form,

$$\log p = a + b\alpha^n + c\beta^n,$$

differentiation gives

$$\frac{1}{p}\frac{dp}{dt} = \frac{}{M^2} b \log \alpha \cdot \alpha^n + \frac{1}{M^2} c \log \beta \cdot \beta^n,$$

in which M is the modulus of the common system of logarithms.

The equation may be written,—

$$\frac{1}{p}\frac{dp}{dt} = A\alpha^n + B\beta^n.$$

The calculation of the values of the constants gives the following results for latitude 45°:—

French units.

B. For 0° to 100° C., mm. of mercury,

$$\log A = 8.8512729 - 10.$$
$$\log B = 6.69305 - 10.$$
$$\log \alpha_1 = 9.996725828 - 10.$$
$$\log \beta_1 = 0.0068641.$$

C. For 100° to 220° C., mm. of mercury.

$$\log A = 8.5495158 - 10.$$
$$\log B = 6.34931 - 10.$$
$$\log \alpha_1 = 9.997411296 - 10.$$
$$\log \beta_1 = 0.0076418.$$

English units.

B. For 32° to 212° F., pounds on the square inch.

$$\log A = 8.5960005 - 10.$$
$$\log B = 6.43778 - 10.$$
$$\log \alpha_2 = 9.998181015 - 10.$$
$$\log \beta_2 = 0.0038134.$$

C. For 212° to 428° F., pounds on the square inch,

$$\log A = 8.2942434 - 10.$$
$$\log B = 6.09403 - 10.$$
$$\log \alpha_2 = 9.998561831 - 10.$$
$$\log \beta_2 = 0.0042454.$$

Heat of the Liquid and Specific Heat. — A preliminary series of experiments convinced Regnault that the specific heat of water at low temperature is unity. To test the specific heat at higher temperatures, he ran hot water from a boiler, and at a known temperature, into a calorimeter in which the temperature varied from 8° to 14° C., and the resulting upper temperature varied from 17° to 29° C. Knowing the original weight of water in the calorimeter, the weight run in from the boiler, and the initial and final temperatures in the calorimeter, he calculated the mean specific heat of water between the temperature of the boiler and the final temperatures of the calorimeter. A series of forty such experiments was made, with the temperature of the boiler varying from 108° to 192° C., from which Regnault concluded that the mean specific heat from 0° to 100° is 1.005; and from 0° to 200°, 1.016. The corresponding heat of the liquid, i.e., the heat required to raise one kilogram of water from 0° to a given temperature, *t*, is

For 100°	100.5
200°	203.2

Assuming an equation of the form

$$q = t + At^2 + Bt^3,$$

and solving for the two constants by aid of the two known values of q, the following equation, which is commonly used, is deduced:

$$q = t + 0.00002t^2 + 0.0000003t^3.$$

The specific heat at any temperature is, therefore, —

$$c = \frac{dq}{dt} = 1 + 0.00004t + 0.000000t^2.$$

These equations are for use with the Centigrade scale; for the Fahrenheit scale, a given temperature may be reduced to the Centigrade scale, and then introduced in the same equations.

The process of making the experiments is really a complex one; for the water, in leaving the boiler, has work done on it by the steam pressure in the boiler, and it has a certain velocity impress on it at the same time, and again, in entering the calorimeter, it does work against the atmospheric pressure, and the kinetic energy of its motion is changed into heat. At higher temperatures there is a double change of state; part of the water changes to steam on leaving the boiler, and that steam is condensed again in the calorimeter. It is probable that the error of neglecting the effect of these several actions is inconsiderable.

The degree of accuracy to be accorded to this work is indicated by the fact that Regnault gives four significant figures in stating the data for the calculation of the constants in the equations.

Rowland's Experiments. — A series of experiments was made by Rowland at Baltimore, to determine the mechanical equivalent of heat, which gave a delicate method of determining the heat of the liquid, and the specific heat.

The apparatus used was similar to that used by Joule, with modifications to give greater certainty of results. The calorimeter was of larger size, and the paddle had the upper vanes curved like the blades of a centrifugal pump, to give a strong circulation up through the centre, past the thermometer for taking the temperatures, and down at the outside. The paddle was driven by a petroleum engine, and the power applied was measured by making the calorimeter into a friction brake, with two arms at which the turning moment was measured. Radiation was made as small as possible, and then was made determinate by use of a water-jacket outside of the calorimeter.

The experiments consisted essentially in delivering a measured amount of work to the water in the calorimeter, and in noting the rise of temperature produced thereby.

The whole range covered by the experiments was from 2° to 41° C. The results show that 430 kilogrammetres of work are required to raise one kilogramme of water from 2° to 3° C. Assuming that the same amount will be required to raise the same weight from 0° to 1° and from 1° to 2°, the following table has been arranged from Rowland's final table of results: —

ROWLAND'S MECHANICAL EQUIVALENT OF HEAT.

Degrees, C.	Total Number of Kilogrammeters.	Mechanical Equivalent of Heat.	Heat of the Liquid, Experimental.	Heat of the Liquid, Calculated.	Degrees, C.	Total Number of Kilogrammeters.	Mechanical Equivalent of Heat.	Heat of the Liquid, Experimental.	Heat of the Liquid, Calculated.
1	430	–	1.0068	1.007	22	9424	426.1	22.065	22.063
2	860	–	2.0135	2.014	23	9850	426.0	23.063	23.061
3	1290	–	3.0204	3.022	24	10277	425.9	24.062	24.059
4	1721	–	4.0295	4.029	25	10701	425.8	25.055	25.058
5	2150	429.8	5.0339	5.036	26	11128	425.7	26.054	26.053
6	2580	429.5	6.0408	6.040	27	11553	425.6	27.050	27.048
7	3009	429.3	7.0452	7.045	28	11978	425.6	28.045	28.042
8	3439	429.0	8.0520	8.049	29	12399	425.5	29.031	29.037
9	3868	428.8	9.0564	9.054	30	12828	425.6	30.035	30.032
10	4296	428.5	10.059	10.058	31	13253	425.6	31.030	31.027
11	4723	428.3	11.058	11.060	32	13675	425.6	32.018	32.023
12	5151	428.1	12.061	12.061	33	14101	425.7	33.016	33.018
13	5578	427.9	13.060	13.063	34	14527	425.7	34.011	34.014
14	6006	427.7	14.063	14.064	35	14952	425.8	35.008	35.009
15	6433	427.4	15.065	15.066	36	15379	425.8	36.008	36.007
16	6861	427.2	16.064	16.066	37	15805	–	37.007	37.005
17	7289	427.0	17.066	17.066	38	16231	–	38.003	38.004
18	7717	426.8	18.068	18.066	39	16657	–	39.000	39.002
19	8144	426.6	19.068	19.066	40	17083	–	39.998	40.000
20	8571	426.4	20.068	20.066	41	17508	–	40.993	
21	8997	426.2	21.065	21.064					

In the above table, column 1 gives the number of degrees above freezing on the Centigrade scale; column 2 gives the number of kilogrammetres required to raise one kilogramme of water from freezing point to the given temperature; column 3 is Rowland's mechanical equivalent of heat at the given temperature derived from 10° intervals on column 2; column 4 is obtained by dividing the numbers in column 2 by the mechanical equivalent of heat at 16⅔° C., or 62° F., from column 3; and column 5 is calculated by considering the specific heat to be constant for each five degrees of temperature. These specific heats were derived from a curve obtained by plotting temperatures for abscissæ, and heats of the liquid for ordinates. The values of the specific heats will be given later, in connection with those for higher temperatures.

A review of the preceding table shows that the specific heat at low temperatures varies quite markedly, so that it appeared advisable to investigate the effect of this variation on Regnault's experiments already quoted. This was done quite expeditiously by multiplying the mean specific heat given by him for his several experiments by the true average specific heat for the range of temperature in the calorimeter. This corrected specific heat was then used to calculate the increase of heat from the final temperature of the calorimeter to the temperature of the boiler, and that increase was added to the heat of the liquid from the table to find the heat of the liquid at the temperature of the boiler. The results were then plotted as before, and compared with the heats of the liquid derived from Regnault's mean specific heats uncorrected. The points by the corrected method were a little more regularly arranged than the points obtained by assuming the specific heat to be unity at low temperatures; but the improvement was inconsiderable. The inequality of the specific heat at low temperatures is seldom so much as the unavoidable errors of the method.

It appeared, that if the specific heat was assumed to be constant, from 40° to 45°, from 45° to 155°, and from 155° to 200° C., the straight lines thus drawn represented the experimental values as recalculated quite nearly; and, further, they represented the uncorrected experimental values more nearly than Regnault's equation.

Specific Heat of Water.—The combination of Rowland's and Regnault's experiments on the heat of the liquid by the method described gives the specific heats set down in the following table, Centigrade scale:—

					SPECIFIC HEAT.
From	0° to	5° C.	32° to	41° F.	1.0072
	5°	10°	41°	50°	1.0044
	10°	15°	50°	59°	1.0016
	15°	20°	59°	68°	1.
	20°	25°	68°	77°	0.9984
	25°	30°	77°	86°	0.9948
	30°	35°	86°	95°	0.9954
	35°	40°	95°	104°	0.9982
	40°	45°	104°	113°	1.
	45°	155°	113°	311°	1.008
	155°	200°	311°	392°	1.046

Thermal Unit.—Heat is measured in calories, or British thermal units (*BTU*). A calorie commonly is defined as the heat required to raise one kilogramme of water from freezing point to 1° C.; and a British thermal unit, that required to raise one pound from 32° to 33° F. Nothing is known about the specific heat of water from 0° to 2° C.; consequently the commonly accepted value of the thermal unit is an ideal quantity inferred from the behavior of water at higher temperatures. It is more scientific to take an easily verified quantity for the standard; and there is a practical convenience in choosing 62° F. for the standard temperature, because it is near the mean temperature of the air during experimental work. Therefore, it is near the mean temperature in the calorimeter during ordinary work with that instrument; and the specific heat of water for the range of temperature in the calorimeter may usually be considered to be unity without error, unless great refinement is desired.

The *BTU* in Tables I and II is taken to be the heat required to raise one pound of water from 62° to 63° F. This agrees substantially with the definition of the calorie, as the heat required to raise one kilogramme of water from 15° to 16° C.

In the tables for other vapors than steam, the old definition for the calorie, and Regnault's value for the heat of the liquid, are retained, to avoid entire recalculation.

Mechanical Equivalent of Heat. — The mechanical equivalent in metre-kilogrammes of one calorie at $16\frac{2}{3}$° C., deduced from Rowland's experiments in the third column of the table on page 58, is 427.1.

Since the value given by Joule is commonly quoted, it will be of interest to make a comparison of his latest work (1873) with Rowland's, as given in the following table: —

Temperature.	Joule's Value at Manchester, English System.	Reduced to the Air Thermometer and to the latitude of Baltimore.		Rowland's Value, corresponding.
		English.	French.	
14.7°	772.7	776.1	425.8	427.6
12.7°	774.6	778.5	427.1	428.0
15.5°	773.1	776.4	426.0	427.3
14.5°	767.0	770.5	422.7	427.5
17.3°	774.0	777.0	426.3	426.9

The value of g at Baltimore, latitude 39° 17′, is 980.05 centimetres therefore, reducing to 45° of latitude, and at the sea level, the value of the mechanical equivalent of heat is

$$J = 426.9.$$

To reduce to the English system, multiply by $\frac{9}{5}$, and by the length of the metre in feet, so that

$$J = 778.$$

Total Heat. — This term is defined as the heat required to raise a unit of weight of water from freezing point to a given temperature, and to entirely evaporate it at that temperature. The experiments made by Regnault were in the reverse order; that is, steam was led from a boiler into the calorimeter, and there condensed. Knowing the initial and final weights of the calorimeter, the temperature of the steam, and the initial and final temperatures of the water in the calorimeter, he was able, after applying the necessary corrections, to calculate the total heats for the several experiments.

As a conclusion of the work, he gives the following values for the total heats: —

10°	610	By equation, 609.6
63°	625	625.2
100°	637	
195°	666	

Assuming an equation of the form

$$\lambda = A + Bt,$$

Regnault calculated the constants from the values given for 100° and 195°, and gives the equation

$$\lambda = 606.5 + 0.305t.$$

Wishing to see the effect of the varying value of the specific heat at low temperatures, I recalculated the total heats given by experiment, by a method resembling that used in recalculation of the heats of the liquid, and plotted the results together with Regnault's values uncorrected. The recalculated points were a little more regular than the original ones, and lay nearer the line represented by the above equation. Especially did the recalculated points for those experiments, for which the true mean specific heat of the water in the calorimeter was nearly unity, lie near that line. It therefore appears that the equation represents our best knowledge of the total heat of steam.

For the Fahrenheit scale the equation becomes

$$\lambda = 1091.7 + 0.305\,(t - 32).$$

Heat of Vaporization.—If the heat of the liquid be subtracted from the total heat, the remainder is called the heat of vaporization, and is represented by r, so that

$$r = \lambda - q.$$

Internal and External Latent Heat. — The heat of vaporization overcomes external pressure, and changes the state from liquid to vapor at constant temperature and pressure. Let the specific volume of the saturated vapor be s, and that of the liquid be σ, then the change of volume is $s - \sigma = u$, on passing from the liquid to the vaporous state. The external work is

$$p(s - \sigma) = pu,$$

and the corresponding amount of heat, or the external latent heat, is

$$Ap(s - \sigma) = Apu,$$

A being the reciprocal of the mechanical equivalent of heat.

The heat required to do the disgregation work, or the internal latent heat, is

$$\rho = r - Apu.$$

Specific Volume and Density of Steam. — On account of the great difficulty of direct determination of the weight of saturated steam, it is customary to calculate the specific volume of steam by aid of the following equation, derived by the application of the principles of thermo-dynamics to the general equation representing the properties of saturated vapor: —

$$s = \frac{r}{AT} \cdot \frac{1}{\frac{dp}{dt}} + \sigma,$$

in which A is the reciprocal of the mechanical equivalent of heat, T is the temperature from the absolute zero, and σ is the volume of one unit of weight of the liquid from which the vapor is formed. The differential co-efficient $\frac{dp}{dt}$ can be calculated by aid of the equations on page 11.

The absolute temperature is obtained by adding 273.7 to the temperature in degrees Centigrade, or 460.7 to the temperature in degrees Fahrenheit.

The volumes and densities of saturated steam given in Tables I, II, and III, were calculated by this method.

It is of interest to consider the degree of accuracy that may be expected from this method of calculating the density of saturated vapor. The value of r repends on λ and q; for the first, Regnault gives three figures in the data from which the empirical equation is deduced, and the experimental work does not indicate a greater degree of accuracy. The fourth figure, if stated, is likely to be in error to the extent of five units. The value of T is commouly stated in four figures, of which the last may be in error by two units. A, as determined by Rowland, has four figures, the last being uncertain to the extent of one or two units. The differential co-efficient $\frac{dp}{dt}$ is deduced from the equations for calculating p; and those equations are derived from data having five places of significant figures. Now the Equations B and C, for steam at 45° of latitude for the English system give a pressure of 14.6967 pounds on the square inch; but the specific volume calculated by aid of Equation B is 26.550 cubic feet, while Equation C gives 26.637 cubic feet. The mean, 26.60, differs from either extreme by about one in seven hundred. This discrepancy is due to the fact that the curves represented by Equations B and C meet at the common temperature, 212°, but do not have a common tangent. Since the equations are empirical and not logical, the error or uncertainty is unavoidable, and all calculated specific volumes are affected by a similar uncertainty. The greatest probable error is in determining r, for which it may be about one in one thousand. The error introduced into this equation by using the values of A in common use, that is, 772 instead of 778, is about one in one hundred.

Tate and Fairbairn's Experiments. — In 1860 an attempt was made by Tate and Fairbairn to determine the specific volume of steam by direct experiment. The following table, taken from the *Philosophical Transactions, Vol. cl.*, gives the results of all their experiments, together with the volumes calculated by their empirical formula,

$$V = 25.62 + \frac{49513}{P + 0.72}.$$

	Pressure in Inches of Mercury. P.	Maximum Temperature, Fahrenheit, of Saturation. T	Specific Volume from Experiments. V.	Specific Volume from Formula. V.	Error of Formula.
1	5.35	136.77	8275.3	8183	$-\frac{1}{90}$
2	8.62	155.33	5333.5	5326	$-\frac{1}{762}$
3	9.45	159.36	4920.2	4900	$-\frac{1}{246}$
4	12.47	170.92	3722.6	3766	$+\frac{1}{87}$
5	12.61	171.48	3715.1	3740	$+\frac{1}{149}$
6	13.62	174.92	3438.1	3478	$+\frac{1}{86}$
7	16.01	182.30	3051.0	2985	$-\frac{1}{46}$
8	18.36	188.30	2623.4	2620	$+\frac{1}{874}$
9	22.88	198.78	2149.5	2124	$-\frac{1}{90}$
1′	53.61	242.90	943.1	937	$-\frac{1}{157}$
2′	55.52	244.82	908.0	906	$-\frac{1}{454}$
3′	55.89	245.22	892.5	900	$+\frac{1}{111}$
4′	66.84	255.50	759.4	758	$-\frac{1}{759}$
5′	76.20	263.14	649.2	669	$+\frac{1}{32}$
6′	81.53	267.21	635.3	628	$-\frac{1}{91}$
7′	84.20	269.20	605.7	608	$+\frac{1}{304}$
8′	92.23	274.76	584.4	562	$-\frac{1}{26}$
9′	90.08	273.30	543.2	545	$+\frac{1}{271}$
10′	99.60	279.42	515.0	519	$+\frac{1}{128}$
11′	104.54	282.58	497.2	496	$-\frac{1}{497}$
12′	112.78	287.25	458.3	461	$+\frac{1}{152}$
13′	122.25	292.53	433.1	428	$-\frac{1}{86}$
14′	114.25	288.25	449.6	456	$+\frac{1}{75}$

It is apparent that the errors of this formula are much larger than the probable errors of the thermo-dynamic method.

The following table, giving the volumes in cubic metres of one kilogramme of saturated steam, shows the comparison of the two methods: —

	0° C.	50° C.	100° C.	150° C.	200° C.
By equation $s = \frac{V}{AT} \cdot \frac{dt}{dp} + \sigma$.	211.5	12.11	1.660	0.3875	0.1277
From equation $V = 25.62 + \frac{49153}{P + 0.72}$,	54.97	11.43	1.643	0.3706	0.1343

Steam Entropy. — From the second law of thermo-dynamics may be deduced the equation

$$d\phi = \frac{dQ}{T},$$

in which ϕ is the entropy, dQ is the heat applied or withdrawn, and T is the absolute temperature. Since the entropy depends on the state of the substance only, and not on the method of arriving at that state, we may calculate the increase of entropy in one unit of weight of a given mixture of water and steam, above the entropy of one pound of water at freezing point, in the following method. Suppose that one unit of weight of water is raised from freezing point to the temperature t, and that the portion x is then changed into steam. During the first operation the change of entropy will be

$$\theta = \int_0 \frac{dq}{T} = \int_0^t \frac{cdt}{T}.$$

During the second operation the change of entropy will be

$$\frac{xr}{T},$$

since the heat is added at the constant temperature t. The entire change of entropy will be

$$\phi = \frac{xr}{T} + \int_0 \frac{cdt}{T} - \frac{xr}{T} + \theta.$$

At any other state the entropy of a unit of weight of a mixture of steam and water will be

$$\phi_1 = \frac{x_1 r_1}{T_1} + \theta_1,$$

and the change of entropy will be

$$\phi - \phi_1 - \frac{xr}{T} + \theta - \frac{x_1 r_1}{T_1} - \theta_1.$$

During an adiabatic change no heat is transmitted, and the entropy is constant.

$$\therefore \quad \frac{xr}{T} + \theta = \frac{xr_1}{T_1} + \theta_1.$$

When the initial state including the value of x is known, and also the final temperature or pressure, the final value of x_1 may be calculated by the above equation; and the initial and final volumes may be found by the equations

$$v = xu + \sigma, \qquad v_1 = x_1 u_1 + \sigma;$$

the value of u for a given temperature or pressure, from the equation,

$$s = u + \sigma.$$

Entropy of the Liquid. — When the specific heat of a liquid is known in terms of the temperature, the entropy of the liquid,

$$\theta = \int_0^t \frac{cdt}{T},$$

is readily calculated. For water we have, for example, the entropy of the liquid at 13° C.

$$1.0072 \log_e \frac{T_5}{T_0} + 1.0044 \log_e \frac{T_{10}}{T_5} + 1.0016 \log_e \frac{T_{13}}{T_{10}}.$$

For other liquids having the general formula for the heat of the liquid,

$$q = at + bt^2 + ct^3,$$

the entropy is

$$\theta = \int_0^t \frac{(a + 2bt + 3ct^2)\,dt}{T}.$$

Other Vapors. — Tables IV to IX are taken from Zenner's *Mechanischen Wärmetheorie*. His values for the specific volume and density were calculated with 273 for the absolute temperature of 0° C., and with 424 for the mechanical equivalent of heat. To bring these tables into accord with Tables I, II, and III, the values of the specific volume and density have been modified by using 273.7 for the absolute temperature of 0° C., and 426.7 for the mechanical equivalent of heat at Paris.

The equations by which the tables were calculated, taken from Regnault's memoirs, *Académie des Sciences*, *Comptes rendus*, *Tome XXXVI*, are here assembled, together with Zeuner's equations for the differential co-efficient, $\frac{1}{p}\frac{dp}{dt}$.

TEMPERATURE AND PRESSURE.

1	log p = 2	a 3	b 4	c 5
Alcohol	$a - ba^n + c\beta^n$	5.4562028	4.9809960	0.0485397
Ether	$a + ba^n - c\beta^n$	5.0286298	0.0002284	3.1906390
Chloroform	$a - ba^n - c\beta^n$	5.2253893	2.9531281	0.0668673
Carbon bisulphide .	$a - ba^n - c\beta^n$	5.4011662	3.4405663	0.2857386
Carbon tetrachloride .	$a - ba^n - c\beta^n$	12.0962331	9.1375180	1.9674890

TEMPERATURE AND PRESSURE — *Concluded.*

	log α. 6	log β. 7	n 8	Limits. 9
Alcohol	$\bar{1}$.99708557	$\bar{1}$.9409485	t+20	−20°, +150°C.
Ether	0.0145775	$\bar{1}$.996877	t+20	−20°, +120°
Chloroform	$\bar{1}$.9974144	$\bar{1}$.9868176	t−20	+20°, +164°
Carbon bisulphide .	1.9977628	$\bar{1}$.9911997	t+20	−20°, +140°
Carbon tetrachloride .	$\bar{1}$.9997120	$\bar{1}$.9949780	t+20	−20°, +188°

The equation for the temperature and pressure of the saturated vapor of aceton, as recalculated by Zeuner, is, —

$$\log p = a - ba^n + c\beta^n.$$
$$a = 5.3085419.$$
$$\log ba^n = + 0.5312766 - 0.0026148t.$$
$$\log c\beta^n = - 0.9645222 - 0.0215592t.$$

$$\frac{1}{p}\frac{dp}{dt} = A\alpha^n + B\beta^n$$

From Zeuner's *Wärmetheorie.*

	Sign.		Log $(A\alpha^n)$	Log $(B\beta^n)$
	$A\alpha^n$	$B\beta^n$		
Alcohol	+	—	$-1.1720041-0.0029143t$	$-2.9992701-0.0590515t$
Ether	+	+	$-1.3396624-0.0031223t$	$-4.4616396+0.0145775t$
Chloroform . . .	+	+	$-1.3410130-0.0025856t$	$-2.0667124-0.0131824t$
Carbon bisulphide .	+	+	$-1.4339778-0.0022372t$	$-2.0511078-0.0088003t$
Carbon tetrachloride,	+	+	$-1.8611078-0.0002880t$	$-1.3812195-0.0050220t$
Aceton	+	+	$-1.3268535-0.0026148t$ t, temperature C.	$-1.9064582-0.0215592t$

HEAT OF THE LIQUID.

Alcohol	$q = 0.54754t + 0.0011218t^2 + 0.000002206t^3$
Ether .	$q = 0.52901t + 0.0002959t^2$
Chloroform .	$q = 0.23235t + 0.0000507t^2$
Carbon bisulphide	$q = 0.23523t + 0.0000815t^2$
Carbon tetrachloride	$q = 0.19798t + 0.0000906t^2$
Aceton . . .	$q = 0.50643t + 0.0003965t^2$

TOTAL HEAT.

Ether . . .	$\lambda = 94 + 0.45t - 0.00055556t^2$
Chloroform .	$\lambda = 67 + 0.1375t$
Carbon bisulphide	$\lambda = 90 + 0.14601t - 0.0004123t^2$
Carbon tetrachloride	$\lambda = 52 + 0.14625t - 0.000172t^2$
Aceton . .	$\lambda = 140.5 + 0.36644t - 0.000516t^2$

The total heat of alcohol varies in so irregular a manner that no equation can be given for it.

Zeuner gives the following empirical equations for calculating the heat equivalent of the internal work, which are proposed to lessen the labor of calculation.

HEAT EQUIVALENT OF INTERNAL WORK.

Water . .	$\rho = 575.40 - 0.791t$
Ether .	$\rho = 86.54 - 0.10648t - 0.0007160t^2$
Chloroform .	$\rho = 62.44 - 0.11282t - 0.0000140t^2$
Carbon bisulphide	$\rho - 82.79 - 0.11446t - 0.0004020t^2$
Carbon tetrachloride .	$\rho = 48.57 - 0.06844t - 0.0002080t^2$
Aceton . . .	$\rho = 131.63 - 0.20184t - 0.0006280t^2$

Sulphur Dioxide and Ammonia. — The use of ice-machines has brought into prominence liquids which vaporize at low temperatures. For two such liquids, sulphur dioxide and ammonia, Regnault gives the following equations for temperature and pressure: —

SULPHUR DIOXIDE.	AMMONIA.
$\log p = a - b\alpha^n - c\beta^n$	$\log p = a - b\alpha^n - c\beta^n$
$a = 5.6663790$	$a = 11.5043330$
$b = 3.0146890$	$b = 7.4503520$
$c = 0.1465400$	$c = 0.9499674$
$\log \alpha = 1.9972989$	$\log \alpha = \bar{1}.9996014$
$\log \beta = 1.9872900$	$\log \beta = 1.9939729$
$n = t + 28$	$n = t + 22$
Limits, -28, $+62$.	Limits, -22, $+82$.

Unfortunately, the heat of the liquid and the total heat for these substances have not been determined. The following approximate method of determining those quantities, and calculating tables, has been proposed by Ledoux, *Annales des Mines*, *1875*.

Zeuner has shown that the properties of superheated steam may be represented by an equation having the form

$$pv = BT - Cp^n,$$

in which

$$B = \frac{c_p n}{A};$$

c_p being the specific heat of superheated steam at constant pressure. C and n are constants to be determined from the properties of the substance. Using 273 for the absolute temperature of freezing point, and 424 for the mechanical equivalent of heat, Zeuner deduces the following values for the constants in the above equation, when the pressure is expressed in kilogrammes per square metre: —

$$B = 50.933 \qquad C = 192.5 \qquad n = \tfrac{4}{3}.$$

The equation has been shown to represent the properties of superheated steam near the point of saturation, and also the properties of dry saturated steam, quite well.

Ledoux assumes that similar equations may be deduced to represent the properties of the vapors of sulphur dioxide and ammonia, founded on Regnault's experiments on the compressibility of gases and vapors, and on the co-efficient of dilatation. Since the co-efficient of dilatation for vapor of ammonia is not given by Regnault, he assumes it to be 0.0039, that for sulphur dioxide being 0.0039028.

The constants deduced for the equations are, —

SULPHUR DIOXIDE.	AMMONIA.
$B = 13.882$	$B = 52.4943$
$C = 3.8455$	$C = 43.7144$
$n = 0.44487$	$n = 0.32685$

These equations he uses for calculating the specific volume and density of the saturated vapors. The total heat is deduced from the adiabatic equations of a superheated vapor. To obtain the heat of vaporization, the equation

$$\frac{r}{u} = AT\frac{dp}{dt}$$

is combined with the equation for a superheated vapor. Thus all the elements are assembled for the calculation of tables of the properties of the saturated vapors. For convenience, empirical equations having the same form as the equations used to represent Regnault's experiments on the total heat and the heat of the liquid are given. They are as follows: —

SULPHUR DIOXIDE.

$$Apu = 8.243 - 0.0196t - 0.000116t^2$$
$$r = 91.396 - 0.2361t - 0.000135t^2$$
$$\lambda = 91.396 + 0.12723t - 0.000131t^2$$
$$q = 0.36333t + 0.00004t^2$$
$$c = 0.36333 + 0.00008t$$

AMMONIA.

$$Apu = 30.154 + 0.08861t - 0.000059t^2$$
$$r = 313.63 - 0.6250t - 0.002111t^2$$
$$\lambda = 313.63 + 0.3808t - 0.000282t^2$$
$$q = 1.0058t + 0.001829t^2$$
$$c = 1.0058 + 0.003658t$$

Finally, Ledoux gives the following equation to represent the pressure in kilogrammes per square metre, to take the place of Regnault's equations quoted on p. 22: —

$$p = a\alpha^{\frac{t}{1-mt}}.$$

SULPHUR DIOXIDE.	AMMONIA.
$a = 15840$	$a = 43474.64$
$\log a = 4.1991752$	$\log a = 4.6382260$
$\alpha = 1.04135$	$\alpha = 1.0386605$
$\log \alpha = 0.0176387$	$\log \alpha = 0.0164736$
$m = 0.0043129$	$m = 0.0040112$

Tables X and XI were calculated by aid of the equations written, and may be of use for approximate calculations, in default of more reliable tables.

Specific Volume of Liquids. — Table XII was taken from the *Phys.-Chem. Tabellen* of Landolt and Börnstein.

Volume of Water. — Table XIII gives the volumes of water compared with its volume at 4°. From 0° to 100° C., the values are those given by Rossetti. Above 100°, the values are those calculated by the equations given by Hirn in the *Annales de Chimie et de Physique, 1867.*

Volumes of Liquids. — The volumes of liquids at high temperatures, compared with the volume of water at 4° C., are represented by the following equations given by Hirn in the *Annales:* —

Liquid	Equation	Logs.
Water 100° C. to 200° C. (vol. at 4° C.= unity)	$v=1+0.00010867875t$	6.0361445−10
	$+0.0000030073653t^2$	4.4781862−10
	$+0.000000028730422t^3$	1.4583419−10
	$-0.0000000000066457031t^4$	8.8225409−20
Alcohol 30° C. to 160° C. (vol. at 0° C.= unity)	$v=1+0.00073892265t$	6.8685991−10
	$+0.00001055235t^2$	3.0233492−10
	$-0.0000000924808 42t^3$	2.9660517−10
	$+0.0000000004041 3567t^4$	0.6065278−10
Ether 30° C. to 130° C. (vol. at 0° C.= unity)	$v=1+0.0013489059t$	7.1299817−10
	$+0.0000065537t^2$	4.8164866−10
	$-0.000000034490756t^3$	2.5377028−10
	$+0.0000000003377 2062t^4$	0.5285571−10
Carbon bisulphide 30° to 160° C. (vol. at 0° C.=unity)	$v=1+0.0011680559t$	7.0674636−10
	$+0.0000016489598t^2$	4.2172103−10
	$-0.0000000081119062t^3$	0.9091229−10
	$+0.000000000060946589t^4$	1.7849494−20
Carbon tetrachloride 30° to 160° C. (vol. at 0°C.=unity)	$v=1+0.0010671883t$	7.0282409−10
	$+0.0000035651378t^2$	4.5520763−10
	$-0.000000014949281t^3$	2.1746202−10
	$+0.000000000085182318t^4$	1.9303494−20

Other Data. — For convenience the following data are assembled: —

Length of the metre in inches	39.3702 (Rogers) 39.370432 (Clarke)
Weight of the kilogramme in pounds . . .	2.20462125
Weight of 1 litre (1 cu. decimetre) of mercury .	13.5959 kilos.
One horse power, in foot pounds per second .	550
Cheval à vapeur, in kilogrammetres per second	75
Normal pressure of the atmosphere .	760 mm. of mercury. 10,333 kilos per sq. m. 14.6967 lbs. per sq. in. 2116.32 lbs. per. sq. ft.
Absolute temperature of freezing point . .	273.°7 C. 492.°7 F.

Explanation of the Tables. — In Table I, the first column gives the temperature, t, of saturated steam.

The second column gives the corresponding pressure, p, in pounds on the square inch, above an absolute vacuum; the differences are placed between the two numbers from which they are derived. For example, the pressure at 40° F. is 0.1216 pounds per square inch; and the difference to be used in interpolation, and placed half a line lower, is 48.

The third column gives the heat of the liquid, q, required to raise the temperature of one pound of water from 32° F. to a given temperature.

The fourth column gives the total heat, λ, required to raise one pound of water from 32° F. to a given temperature, and to entirely vaporize it under the pressure due to that temperature.

The fifth column gives the heat of vaporization, or the heat required to vaporize one pound of water at a given temperature, under the pressure corresponding.

The sixth column gives the heat required to do the disgregation work during the vaporization of one pound of water.

The seventh column gives the heat required to overcome the external pressure, and do the work of increasing the volume from σ to s.

The eighth column gives the entropy of the liquid.

The ninth and tenth columns give the specific volume, or volume in cubic feet, of one pound of saturated steam, and the density or weight of one cubic foot in pounds.

Table II differs from Table I in that it is arranged to give the properties of saturated steam for each pound of pressure.

Table III gives the properties of saturated steam in French units; and Tables IV to XI give the properties of other saturated vapors in the same units. It is to be noted that the pressures in Tables IV to IX are in millimetres of mercury; in Tables X and XI, are in kilogrammes per square metre.

TABLE I.

SATURATED STEAM.

ENGLISH UNITS.

Temperature, Degrees Fahr. t	Pressure, Pounds per Square Inch. p		Heat of the Liquid. q	Total Heat. λ	Heat of Vaporization. r	Heat equivalent of Internal Work. ρ	Heat equivalent of External Work. Apu	Entropy of the Liquid. $\int \frac{cdt}{T}$	Specific Volume. s		Density. Weight, in Pounds, of one Cubic Foot. γ		Temperature, Degrees Fahr. t
32	0.0890	36	0	1091.7	1091.7	1035.9	55.8	0.0000	3387	127	0.0002952	115	**32**
33	0.0926	37	1.01	1092.0	1091.0	1035.1	55.9	0.0020	3260	122	0.0003067	120	**33**
34	0.0963	39	2.01	1092.3	1090.3	1034.3	56.0	0.0041	3138	116	0.0003187	122	**34**
35	0.1002	40	3.02	1092.6	1089.6	1033.6	56.0	0.0061	3022	112	0.0003309	127	**35**
36	0.1042	41	4.03	1092.9	1088.9	1032.8	56.1	0.0081	2910	107	0.0003436	132	**36**
37	0.1083	43	5.04	1093.2	1088.2	1032.0	56.2	0.0101	2803	103	0.0003568	136	**37**
38	0.1126	44	6.04	1093.5	1087.5	1031.3	56.2	0.0122	2700	99	0.0003704	141	**38**
39	0.1170	46	7.05	1093.8	1086.7	1030.4	56.3	0.0142	2601	95	0.0003845	145	**39**
40	0.1216	48	8.06	1094.1	1086.0	1029.6	56.4	0.0162	2506	91	0.0003990	151	**40**
41	0.1264	49	9.06	1094.4	1085.3	1028.8	56.5	0.0182	2415	87	0.0004141	155	**41**
42	0.1313	51	10.07	1094.8	1084.7	1028.1	56.6	0.0202	2328	84	0.0004296	160	**42**
43	0.1364	53	11.07	1095.1	1084.0	1027.3	56.7	0.0222	2244	80	0.0004456	165	**43**
44	0.1417	54	12.08	1095.4	1083.3	1026.5	56.8	0.0242	2164	77	0.0004621	171	**44**
45	0.1471	57	13.08	1095.7	1082.6	1025.8	56.8	0.0262	2087	74	0.0004792	176	**45**
46	0.1528	58	14.09	1096.0	1081.9	1025.0	56.9	0.0282	2013	71	0.0004968	181	**46**
47	0.1586	60	15.09	1096.3	1081.2	1024.2	57.0	0.0302	1942	68	0.0005149	187	**47**
48	0.1646	62	16.10	1096.6	1080.5	1023.4	57.1	0.0322	1874	66	0.0005336	194	**48**
49	0.1708	65	17.10	1096.9	1079.8	1022.6	57.2	0.0341	1808	63	0.0005530	201	**49**
50	0.1773	66	18.10	1097.2	1079.1	1021.8	57.3	0.0361	1745	60	0.0005731	206	**50**
51	0.1839	69	19.11	1097.5	1078.4	1021.1	57.3	0.0381	1685	59	0.0005937	213	**51**
52	0.1908	71	20.11	1097.8	1077.7	1020.3	57.4	0.0400	1626	56	0.0006150	219	**52**
53	0.1979	73	21.11	1098.1	1077.0	1019.5	57.5	0.0420	1570	54	0.0006369	226	**53**
54	0.2052	76	22.11	1098.4	1076.3	1018.7	57.6	0.0439	1516	51	0.0006595	234	**54**
55	0.2128	78	23.11	1098.7	1075.6	1017.9	57.7	0.0459	1465	50	0.0006829	240	**55**
56	0.2206	81	24.11	1099.0	1074.9	1017.1	57.8	0.0478	1415	48	0.0007069	248	**56**
57	0.2287	83	25.12	1099.3	1074.2	1016.3	57.9	0.0497	1367	46	0.0007317	254	**57**
58	0.2370	86	26.12	1099.6	1073.5	1015.6	57.9	0.0517	1321	45	0.0007571	263	**58**
59	0.2456	89	27.12	1099.9	1072.8	1014.8	58.0	0.0536	1276	42	0.0007834	270	**59**
60	0.2545	92	28.12	1100.2	1072.1	1014.0	58.1	0.0555	1234	41	0.0008104	280	**60**
61	0.2637	94	29.12	1100.5	1071.4	1013.2	58.2	0.0574	1193	40	0.0008384	289	**61**
62	0.2731	98	30.12	1100.9	1070.8	1012.5	58.3	0.0594	1153	38	0.0008673	296	**62**
63	0.2829	100	31.12	1101.2	1070.1	1011.7	58.4	0.0613	1115	37	0.0008969	304	**63**

SATURATED STEAM — *Continued.*

Temperature, Degrees Fahr. t	Pressure, Pounds per Square Inch. p	Heat of the Liquid. q	Total Heat. λ	Heat of Vaporization. r	Heat equivalent of Internal Work. ρ	Heat equivalent of External Work. Apu	Entropy of the Liquid. $\int \frac{cdt}{T}$	Specific Volume. s	DENSITY. Weight, in Pounds, of one Cubic Foot. γ	Temperature, Degrees Fahr. t
64	0.2929$_{104}$	32.12	1101.5	1069.4	1010.9	58.5	0.0632	1078$_{36}$	0.0009273$_{313}$	**64**
65	0.3033$_{107}$	33.12	1101.8	1068.7	1010.1	58.6	0.0651	1042$_{33}$	0.0009586$_{325}$	**65**
66	0.3140$_{110}$	34.12	1102.1	1068.0	1009.4	58.6	0.0670	1009$_{33}$	0.0009911$_{329}$	**66**
67	0.3250$_{114}$	35.12	1102.4	1067.3	1008.6	58.7	0.0689	976.3$_{316}$	0.001024$_{35}$	**67**
68	0.3364$_{117}$	36.12	1102.7	1066.6	1007.8	58.8	0.0708	944.7$_{304}$	0.001059$_{35}$	**68**
69	0.3481$_{121}$	37.12	1103.0	1065.9	1007.0	58.9	0.0727	914.3$_{293}$	0.001094$_{36}$	**69**
70	0.3602$_{124}$	38.11	1103.3	1065.2	1006.2	59.0	0.0745	885.0$_{283}$	0.001130$_{37}$	**70**
71	0.3726$_{128}$	39.11	1103.6	1064.5	1005.4	59.1	0.0764	856.7$_{272}$	0.001167$_{38}$	**71**
72	0.3854$_{132}$	40.11	1103.9	1063.8	1004.6	59.2	0.0783	829.5$_{263}$	0.001205$_{40}$	**72**
73	0.3986$_{136}$	41.11	1104.2	1063.1	1003.8	59.3	0.0802	803.2$_{253}$	0.001245$_{41}$	**73**
74	0.4122$_{140}$	42.11	1104.5	1062.4	1003.0	59.4	0.0820	777.9$_{244}$	0.001286$_{41}$	**74**
75	0.4262$_{144}$	43.11	1104.8	1061.7	1002.3	59.4	0.0839	753.5$_{236}$	0.001327$_{43}$	**75**
76	0.4406$_{149}$	44.11	1105.1	1061.0	1001.5	59.5	0.0858	729.9$_{228}$	0.001370$_{44}$	**76**
77	0.4555$_{153}$	45.10	1105.4	1060.3	1000.7	59.6	0.0876	707.1$_{219}$	0.001414$_{45}$	**77**
78	0.4708$_{157}$	46.10	1105.7	1059.6	999.9	59.7	0.0895	685.2$_{211}$	0.001459$_{46}$	**78**
79	0.4865$_{162}$	47.09	1106.0	1058.9	999.1	59.8	0.0913	664.1$_{203}$	0.001505$_{48}$	**79**
80	0.5027$_{167}$	48.09	1106.3	1058.2	998.3	59.9	0.0932	643.8$_{197}$	0.001553$_{49}$	**80**
81	0.5194$_{171}$	49.08	1106.6	1057.5	997.5	60.0	0.0950	624.1$_{191}$	0.001602$_{51}$	**81**
82	0.5365$_{177}$	50.08	1107.0	1056.9	996.8	60.1	0.0968	605.0$_{184}$	0.001653$_{52}$	**82**
83	0.5542$_{181}$	51.07	1107.3	1056.2	996.0	60.2	0.0987	586.6$_{178}$	0.001705$_{53}$	**83**
84	0.5723$_{187}$	52.07	1107.6	1055.5	995.2	60.3	0.1005	568.8$_{171}$	0.001758$_{55}$	**84**
85	0.5910$_{192}$	53.06	1107.9	1054.8	994.4	60.4	0.1023	551.7$_{165}$	0.001813$_{56}$	**85**
86	0,6102$_{197}$	54.06	1108.2	1054.1	993.7	60.4	0.1041	535.2$_{160}$	0.001869$_{57}$	**86**
87	0.6299$_{203}$	55.05	1108.5	1053.4	992.9	60.5	0.1060	519.2$_{155}$	0.001926$_{59}$	**87**
88	0.6502$_{209}$	56.05	1108.8	1052.7	992.1	60.6	0.1078	503.7$_{148}$	0.001985$_{60}$	**88**
89	0.6711$_{214}$	57.04	1109.1	1052.1	991.4	60.7	0.1096	488.9$_{143}$	0.002045$_{62}$	**89**
90	0.6925$_{221}$	58.04	1109.4	1051.4	990.6	60.8	0.1114	474.6$_{139}$	0.002107$_{64}$	**90**
91	0.7146$_{226}$	59.03	1109.7	1050.7	989.8	60.9	0.1132	460.7$_{136}$	0.002171$_{66}$	**91**
92	0.7372$_{233}$	60.03	1110.0	1050.0	989.0	61.0	0.1150	447.1$_{131}$	0.002237$_{67}$	**92**
93	0.7605$_{239}$	61.03	1110.3	1049.3	988.2	61.1	0.1168	434.0$_{125}$	0.002304$_{68}$	**93**
94	0.7844$_{246}$	62.02	1110.6	1048.6	987.4	61.2	0.1186	421.5$_{122}$	0.002372$_{71}$	**94**
95	0.8090$_{252}$	63.02	1110.9	1047.9	986.6	61.3	0.1204	409.3$_{118}$	0 002443$_{73}$	**95**
96	0.8342$_{259}$	64.01	1111.2	1047.2	985.8	61.4	0.1222	397.5$_{114}$	0.002516$_{74}$	**96**
97	0.8601$_{266}$	65.01	1111.5	1046.5	985.0	61.5	0.1240	386.1$_{110}$	0.002590$_{76}$	**97**
98	0.8867$_{273}$	66.01	1111.8	1045.8	984.2	61.6	0.1258	375.1$_{107}$	0.002666$_{78}$	**98**
99	0.9140$_{281}$	67.01	1112.1	1045.1	983.4	61.7	0.1275	364.4$_{104}$	0.002744$_{80}$	**99**
100	0.9421$_{288}$	68.01	1112.4	1044.4	982.7	61.7	0.1293	354.0$_{99}$	0.002824$_{82}$	**100**
101	0.9709$_{295}$	69.01	1112.7	1043.7	981.9	61.8	0.1311	344.1$_{96}$	0.002906$_{84}$	**101**
102	1.0004$_{303}$	70.00	1113.1	1043.1	981.2	61.9	0.1329	334.5$_{93}$	0.002990$_{85}$	**102**
103	1.0307$_{312}$	71.00	1113.4	1042.4	980.4	62.0	0.1347	325.2$_{91}$	0.003075$_{88}$	**103**

SATURATED STEAM — *Continued.*

Temperature, Degrees Fahr. t	Pressure, Pounds per Square Inch. p	Heat of the Liquid. q	Total Heat. λ	Heat of Vaporization. r	Heat equivalent of Internal Work. ρ	Heat equivalent of External Work. Apu	Entropy of the Liquid. $\int \frac{cdt}{T}$	Specific Volume. s	Density. Weight, in Pounds, of one Cubic Foot. γ	Temperature, Degrees Fahr. t
104	1.0619_{319}	72.0	1113.7	1041.7	979.6	62.1	0.1364	316.1_{88}	0.003163_{91}	**104**
105	1.0938_{328}	73.0	1114.0	1041.0	978.8	62.2	0.1382	307.3_{85}	0.003254_{93}	**105**
106	1.1266_{336}	74.0	1114.3	1040.3	978.0	62.3	0.1400	298.8_{82}	0.003347_{94}	**106**
107	1.1602_{345}	75.0	1114.6	1039.6	977.2	62.4	0.1417	290.6_{79}	0.003441_{96}	**107**
108	1.1947_{354}	76.0	1114.9	1038.9	976.4	62.5	0.1435	282.7_{77}	0.003537_{99}	**108**
109	1.2301_{362}	77.0	1115.2	1038.2	975.6	62.6	0.1452	275.0_{75}	0.003636_{102}	**109**
110	1.2663_{372}	78.0	1115.5	1037.5	974.8	62.7	0.1470	267.5_{72}	0.003738_{104}	**110**
111	1.3035_{381}	79.0	1115.8	1036.8	974.0	62.8	0.1487	260.3_{70}	0.003842_{106}	**111**
112	1.3416_{391}	80.0	1116.1	1036.1	973.2	62.9	0.1505	253.3_{68}	0.003948_{109}	**112**
113	1.3807_{400}	81.0	1116.4	1035.4	972.4	63.0	0.1522	246.5_{66}	0.004057_{111}	**113**
114	1.4207_{411}	82.0	1116.7	1034.7	971.6	63.1	0.1540	239.9_{64}	0.004168_{115}	**114**
115	1.4618_{421}	83.0	1117.0	1034.0	970.8	63.2	0.1558	233.5_{62}	0.004283_{116}	**115**
116	1.5039_{431}	84.0	1117.3	1033.3	970.0	63.3	0.1575	227.3_{60}	0.004399_{120}	**116**
117	1.5470_{442}	85.0	1117.6	1032.6	969.2	63.4	0.1592	221.3_{58}	0.004519_{121}	**117**
118	1.5912_{452}	86.0	1117.9	1031.9	968.4	63.5	0.1610	215.5_{56}	0.004640_{124}	**118**
119	1.6364_{464}	87.0	1118.2	1031.2	967.6	63.6	0.1627	209.9_{55}	0.004764_{128}	**119**
120	1.6828_{474}	88.1	1118.5	1030.4	966.7	63.7	0.1645	204.4_{53}	0.004892_{130}	**120**
121	1.7302_{487}	89.1	1118.8	1029.7	966.0	63.7	0.1662	199.1_{52}	0.005022_{134}	**121**
122	1.7789_{498}	90.1	1119.2	1029.1	965.3	63.8	0.1679	193.9_{50}	0.005156_{137}	**122**
123	1.8287_{510}	91.1	1119.5	1028.4	964.5	63.9	0.1697	188.9_{48}	0.005293_{139}	**123**
124	1.8797_{521}	92.1	1119.8	1027.7	963.7	64.0	0.1714	184.1_{47}	0.005432_{142}	**124**
125	1.9318_{534}	93.1	1120.1	1027.0	962.9	64.1	0.1731	179.4_{46}	0.005574_{146}	**125**
126	1.9852_{547}	94.1	1120.4	1026.3	962.1	64.2	0.1748	174.8_{44}	0.005720_{148}	**126**
127	2.0399_{560}	95.1	1120.7	1025.6	961.3	64.3	0.1765	170.4_{43}	0.005868_{152}	**127**
128	2.0959_{574}	96.1	1121.0	1024.9	960.5	64.4	0.1783	166.1_{42}	0.006020_{156}	**128**
129	2.1533_{586}	97.1	1121.3	1024.2	959.7	64.5	0.1800	161.9_{41}	0.006176_{160}	**129**
130	2.2119_{600}	98.1	1121.6	1023.5	958.9	64.6	0.1817	157.8_{39}	0.006336_{162}	**130**
131	2.2719_{614}	99.1	1121.9	1022.8	958.1	64.7	0.1834	153.9_{38}	0.006498_{166}	**131**
132	2.3333_{628}	100.2	1122.2	1022.0	957.2	64.8	0.1851	150.1_{37}	0.006664_{169}	**132**
133	2.3961_{642}	101.2	1122.5	1021.3	956.4	64.9	0.1868	146.4_{36}	0.006833_{172}	**133**
134	2.4603_{658}	102.2	1122.8	1020.6	955.6	65.0	0.1885	142.8_{36}	0.007005_{176}	**134**
135	2.5261_{671}	103.2	1123.1	1019.9	954.8	65.1	0.1902	139.2_{34}	0.007181_{180}	**135**
136	2.5932_{687}	104.2	1123.4	1019.2	954.0	65.2	0.1919	135.8_{33}	0.007361_{184}	**136**
137	2.6619_{702}	105.2	1123.7	1018.5	953.2	65.3	0.1936	132.5_{32}	0.007545_{187}	**137**
138	2.7321_{719}	106.2	1124.0	1017.8	952.4	65.4	0.1952	129.3_{31}	0.007732_{192}	**138**
139	2.8040_{734}	107.2	1124.3	1017.1	951.6	65.5	0.1969	126.2_{30}	0.007924_{196}	**139**
140	2.8774_{751}	108.2	1124.6	1016.4	950.8	65.6	0.1986	123.2_{30}	0.008120_{198}	**140**
141	2.9525_{767}	109.2	1124.9	1015.7	950.0	65.7	0.2003	120.2_{29}	0.008318_{204}	**141**
142	3.0292_{784}	110.2	1125.3	1015.1	949.3	65.8	0.2020	117.3_{28}	0.008522_{208}	**142**
143	3.1076_{801}	111.2	1125.6	1014.4	948.5	65.9	0.2036	114.5_{27}	0.008730_{212}	**143**

SATURATED STEAM—*Continued.*

Temperature, Degrees Fahr. t	Pressure, Pounds per Square Inch. p	Heat of the Liquid. q	Total Heat. λ	Heat of Vaporization. r	Heat equivalent of Internal Work. ρ	Heat equivalent of External Work. Apu	Entropy of the Liquid. $\int\frac{cdt}{T}$	Specific Volume. s	Density. Weight, in Pounds, of one Cubic Foot. γ	Temperature, Degrees Fahr. t
144	3.1877_{819}	112.2	1125.9	1013.7	947.7	66.0	0.2053	111.8_{26}	0.008942_{217}	**144**
145	3.2696_{836}	113.3	1126.2	1012.9	946.8	66.1	0.2070	109.2_{26}	0.009159_{220}	**145**
146	3.3532_{855}	114.3	1126.5	1012.2	946.0	66.2	0.2086	106.6_{25}	0.009379_{225}	**146**
147	3.4387_{873}	115.3	1126.8	1011.5	945.2	66.3	0.2103	104.1_{24}	0.009604_{229}	**147**
148	3.5260_{892}	116.3	1127.1	1010.8	944.4	66.4	0.2119	101.7_{24}	0.009833_{237}	**148**
149	3.6152_{911}	117.3	1127.4	1010.1	943.6	66.5	0.2136	99.33_{230}	0.01007_{24}	**149**
150	3.7063_{930}	118.3	1127.7	1009.4	942.8	66.6	0.2152	97.03_{224}	0.01031_{24}	**150**
151	3.7993_{950}	119.3	1128.0	1008.7	942.0	66.7	0.2169	94.79_{218}	0.01055_{25}	**151**
152	3.8943_{970}	120.3	1128.3	1008.0	941.3	66.7	0.2185	92.61_{212}	0.01080_{25}	**152**
153	3.9913_{990}	121.3	1128.6	1007.3	940.5	66.8	0.2202	90.49_{206}	0.01105_{26}	**153**
154	4.0903_{1011}	122.3	1128.9	1006.6	939.7	66.9	0.2218	88.43_{201}	0.01131_{26}	**154**
155	4.1914_{1032}	123.3	1129.2	1005.9	938.9	67.0	0.2235	86.42_{195}	0.01157_{27}	**155**
156	4.2946_{1054}	124.3	1129.5	1005.2	938.1	67.1	0.2251	84.47_{191}	0.01184_{27}	**156**
157	4.4000_{1075}	125.4	1129.8	1004.4	937.2	67.2	0.2267	82.56_{186}	0.01211_{28}	**157**
158	4.5075_{1097}	126.4	1130.1	1003.7	936.4	67.3	0.2284	80.70_{180}	0.01239_{28}	**158**
159	4.6172_{1120}	127.4	1130.4	1003.0	935.6	67.4	0.2300	78.90_{176}	0.01267_{29}	**159**
160	4.7292_{1143}	128.4	1130.7	1002.3	934.8	67.5	0.2316	77.14_{171}	0.01296_{30}	**160**
161	4.8435_{1166}	129.4	1131.0	1001.6	934.0	67.6	0.2332	75.43_{166}	0.01326_{30}	**161**
162	4.9601_{1189}	130.4	1131.4	1001.0	933.3	67.7	0.2349	73.77_{163}	0.01356_{30}	**162**
163	5.079_{121}	131.4	1131.7	1000.3	932.5	67.8	0.2365	72.14_{158}	9.01386_{31}	**163**
164	5.200_{124}	132.4	1132.0	999.6	931.7	67.9	0.2381	70.56_{155}	0.01417_{32}	**164**
165	5.324_{126}	133.4	1132.3	998.9	930.9	68.0	0.2397	69.01_{150}	0.01449_{32}	**165**
166	5.450_{129}	134.4	1132.6	998.2	930.1	68.1	0.2413	67.51_{146}	0.01481_{33}	**166**
167	5.579_{131}	135.4	1132.9	997.5	929.3	68.2	0.2429	66.05_{143}	0.01514_{34}	**167**
168	5.710_{134}	136.4	1133.2	996.8	928.5	68.3	0.2445	64.62_{140}	0.01548_{34}	**168**
169	5.844_{137}	137.4	1133.5	996.1	927.7	68.4	0.2461	63.22_{137}	0.01582_{35}	**169**
170	5.981_{139}	138.5	1133.8	995.3	926.8	68.5	0.2477	61.85_{132}	0.01617_{35}	**170**
171	6.120_{142}	139.5	1134.1	994.6	926.0	68.6	0.2493	60.53_{128}	0.01652_{36}	**171**
172	6.262_{145}	140.5	1134.4	993.9	925.2	68.7	0.2509	59.25_{126}	0.01688_{36}	**172**
173	6.407_{147}	141.5	1134.7	993.2	924.4	68.8	0.2525	57.99_{123}	0.01724_{38}	**173**
174	6.554_{150}	142.5	1135.0	992.5	923.7	68.8	0.2541	56.76_{120}	0.01762_{38}	**174**
175	6.704_{154}	143.5	1135.3	991.8	922.9	68.9	0.2557	55.56_{116}	0.01800_{38}	**175**
176	6.858_{156}	144.5	1135.6	991.1	922.1	69.0	9.2573	54.40_{114}	0.01838_{40}	**176**
177	7.014_{159}	145.5	1135.9	990.4	921.3	69.1	0.2589	53.26_{112}	0.01878_{40}	**177**
178	7.173_{162}	146.5	1136.2	989.7	920.5	69.2	0.2604	52.14_{108}	0.01918_{40}	**178**
179	7.335_{165}	147.5	1136.5	989.0	919.7	69.3	0.2620	51.06_{105}	0.01958_{42}	**179**
180	7.500_{168}	148.5	1136.8	988.3	918.9	69.4	0.2636	50.01_{103}	0.02000_{42}	**180**
181	7.668_{172}	149.5	1137.1	987.6	918.1	69.5	0.2652	48.98_{101}	0.02042_{43}	**181**
182	7.840_{174}	150.6	1137.5	986.9	917.3	69.6	0.2667	47.97_{98}	0.02085_{43}	**182**
183	8.014_{178}	151.6	1137.8	986.2	916.5	69.7	0.2683	46.99_{96}	0.02128_{44}	**183**

SATURATED STEAM—*Continued.*

Temperature, Degrees Fahr. t	Pressure, Pounds per Square Inch. p	Heat of the Liquid. q	Total Heat. λ	Heat of Vaporization. r	Heat equivalent of Internal Work. ρ	Heat equivalent of External Work. Apu	Entropy of the Liquid. $\int \frac{cdt}{T}$	Specific Volume. s	Density. Weight, in Pounds, of one Cubic Foot. γ	Temperature, Degrees Fahr. t
184	8.192_{181}	152.6	1138.1	985.5	915.7	69.8	0.2699	46.03_{94}	0.02172_{46}	**184**
185	8.373_{185}	153.6	1138.4	984.8	914.9	69.9	0.2714	45.09_{92}	0.02218_{46}	**185**
186	8.558_{188}	154.6	1138.7	984.1	914.1	70.0	0.2730	44.17_{89}	0.02264_{47}	**186**
187	8.746_{191}	155.6	1139.0	983.4	913.4	70.0	0.2745	43.28_{87}	0.02311_{47}	**187**
188	8.937_{195}	156.6	1139.3	982.7	912.6	70.1	0.2761	42.41_{85}	0.02358_{48}	**188**
189	9.132_{198}	157.6	1139.6	982.0	901.8	70.2	0.2777	41.56_{83}	0.02406_{49}	**189**
190	9.330_{202}	158.6	1139.9	981.3	911.0	70.3	0.2792	40.73_{81}	0.02455_{50}	**190**
191	9.532_{206}	159.6	1140.2	980.6	910.2	70.4	0.2808	39.92_{79}	0.02505_{51}	**191**
192	9.738_{209}	160.6	1140.5	979.9	909.4	70.5	0.2823	39.13_{78}	0.02556_{52}	**192**
193	9.947_{213}	161.6	1140.8	979.2	908.6	70.6	0.2838	38.35_{76}	0.02608_{52}	**193**
194	10.160_{217}	162.6	1141.1	978.5	907.8	70.7	0.2854	37.59_{74}	0.02660_{54}	**194**
195	10.377_{221}	163.7	1141.4	977.7	906.9	70.8	0.2869	36.85_{72}	0.02714_{54}	**195**
196	10.598_{224}	164.7	1141.7	977.0	906.2	70.8	0.2885	36.13_{71}	0.02768_{55}	**196**
197	10.822_{229}	165.7	1142.0	976.3	905.4	70.9	0.2900	35.42_{69}	0.02823_{56}	**197**
198	11.051_{232}	166.7	1142.3	975.6	904.6	71.0	0.2915	34.73_{67}	0.02879_{57}	**198**
199	11.283_{237}	167.7	1142.6	974.9	903.8	71.1	0.2930	34.06_{66}	0.02936_{58}	**199**
200	11.520_{241}	168.7	1142.9	974.2	903.0	71.2	0.2946	33.40_{64}	0.02994_{59}	**200**
201	11.761_{244}	169.7	1143.2	973.5	902.2	71.3	0.2961	32.76_{63}	0.03053_{59}	**201**
202	12.005_{249}	170.7	1143.6	972.9	901.5	71.4	0.2976	32.13_{61}	0.03112_{61}	**202**
203	12.254_{254}	171.7	1143.9	972.2	900.8	71.4	0.2991	31.52_{60}	0.03173_{62}	**203**
204	12.508_{257}	172.7	1144.2	971.5	900.0	71.5	0.3007	30.92_{59}	0.03235_{62}	**204**
205	12.765_{263}	173.7	1144.5	970.8	899.2	71.6	0.3022	30.33_{57}	0.03297_{66}	**205**
206	13.028_{266}	174.7	1144.8	970.1	898.4	71.7	0.3037	29.76_{57}	0.03361_{65}	**206**
207	13.294_{271}	175.8	1145.1	969.3	897.5	71.8	0.3052	29.19_{56}	0.03426_{67}	**207**
208	13.565_{276}	176.8	1145.4	968.6	896.7	71.9	0.3067	28.63_{54}	0.03493_{67}	**208**
209	13.841_{281}	177.8	1145.7	967.9	896.0	71.9	0.3082	28.09_{52}	0.03560_{68}	**209**
210	14.122_{285}	178.8	1146.0	967.2	895.2	72.0	0.3097	27.57_{52}	0.03628_{69}	**210**
211	14.407_{290}	179.8	1146.3	966.5	894.4	72.1	0.3112	27.05_{45}	0.03697_{63}	**211**
212	14.697_{293}	180.8	1146.6	965.8	893.5	72.3	0.3127	26.60_{44}	0.03760_{64}	**212**
213	14.990_{299}	181.8	1146.9	965.1	892.6	72.5	0.3142	26.16_{49}	0.03824_{72}	**213**
214	15.289_{303}	182.8	1147.2	964.4	891.8	72.6	0.3157	25.67_{48}	0.03896_{73}	**214**
215	15.592_{309}	183.8	1147.5	963.7	891.0	72.7	0.3172	25.19_{46}	0.03969_{74}	**215**
216	15.901_{313}	184.8	1147.8	963.0	890.2	72.8	0.3187	24.73_{45}	0.04043_{75}	**216**
217	16.214_{319}	185.8	1148.1	962.3	889.5	72.8	0.3202	24.28_{44}	0.04118_{76}	**217**
218	16.533_{324}	186.8	1148.4	961.6	888.7	72.9	0.3217	23.84_{43}	0.04194_{78}	**218**
219	16.857_{329}	187.8	1148.7	960.9	887.9	73.0	0.3232	23.41_{43}	0.04272_{80}	**219**
220	17.186_{335}	188.9	1149.0	960.1	887.1	73.0	0.3246	22.98_{42}	0.04352_{80}	**220**
221	17.521_{340}	189.9	1149.3	959.4	886.3	73.1	0.3261	22.56_{41}	0.04432_{82}	**221**
222	17.861_{345}	190.9	1149.7	958.8	885.6	73.2	0.3276	22.15_{39}	0.04514_{82}	**222**
223	18.206_{351}	191.9	1150.0	958.1	884.8	73.3	0.3291	21.76_{39}	0.04596_{84}	**223**

SATURATED STEAM — *Continued.*

Temperature, Degrees Fahr. t	Pressure, Pounds per Square Inch. p	Heat of the Liquid. q	Total Heat. λ	Heat of Vaporization. r	Heat equivalent of Internal Work. ρ	Heat equivalent of External Work. Apu	Entropy of the Liquid. $\int \frac{cdt}{T}$	Specific Volume. s	Density. Weight, in Pounds, of one Cubic Foot. γ	Temperature, Degrees Fahr. t
224	18.557_{357}	192.9	1150.3	957.4	884.0	73.4	0.3305	21.37_{38}	0.04679_{85}	**224**
225	18.914_{362}	193.9	1150.6	956.7	883.3	73.4	0.3320	20.99_{37}	0.04764_{86}	**225**
226	19.276_{368}	194.9	1150.9	956.0	882.5	73.5	0.3335	20.62_{37}	0.04850_{88}	**226**
227	19.644_{374}	195.9	1151.2	955.3	881.7	73.6	0.3349	20.25_{36}	0.04938_{90}	**227**
228	20.018_{379}	196.9	1151.5	954.6	880.9	73.7	0.3364	19.89_{35}	0.05028_{90}	**228**
229	20.397_{386}	197.9	1151.8	953.9	880.2	73.7	0.3379	19.54_{34}	0.05118_{90}	**229**
230	20.783_{392}	198.9	1152.1	953.2	879.4	73.8	0.3393	19.20_{33}	0.05208_{92}	**230**
231	21.175_{397}	199.9	1152.4	952.5	878.6	73.9	0.3408	18.87_{33}	0.05300_{94}	**231**
232	21.572_{404}	201.0	1152.7	951.7	877.8	73.9	0.3423	18.54_{32}	0.65394_{95}	**232**
233	21.976_{410}	202.0	1153.0	951.0	877.0	74.0	0.3437	18.22_{32}	0.05489_{97}	**233**
234	22.386_{417}	203.0	1153.3	950.3	876.2	74.1	0.3452	17.90_{31}	0.05586_{99}	**234**
225	22.803_{423}	204.0	1153.6	949.6	875.4	74.2	0.3466	17.59_{30}	0.05685_{99}	**235**
236	23.226_{429}	205.0	1153.9	948.9	874.6	74.3	0.3481	17.29_{30}	0.05784_{101}	**236**
237	23.655_{436}	206.0	1154.2	948.2	873.9	74.3	0.3495	16.99_{29}	0.05885_{102}	**237**
238	24.091_{442}	207.0	1154.5	947.5	873.1	74.4	0.3510	16.70_{28}	0.05987_{103}	**238**
239	24.533_{449}	208.0	1154.8	946.8	872.3	74.5	0.3524	16.42_{28}	0.06090_{105}	**239**
240	24.982_{456}	209.0	1155.1	946.1	871.6	74.5	0.3538	16.14_{27}	0.06195_{106}	**240**
241	25.438_{462}	210.0	1155.4	945.4	870.8	74.6	0.3553	15.87_{27}	0.06301_{108}	**241**
242	25.900_{470}	211.0	1155.8	944.8	870.1	74.7	0.3567	15.60_{26}	0.06409_{110}	**242**
243	26.370_{476}	212.0	1156.1	944.1	869.3	74.8	0.3581	15.34_{26}	0.06519_{111}	**243**
244	26.846_{484}	213.0	1156.4	943.4	868.5	74.9	0.3596	15.08_{25}	0.06630_{113}	**244**
245	27.330_{491}	214.1	1156.7	942.6	867.7	74.9	0.3610	14.83_{25}	0.06743_{115}	**245**
246	27.821_{498}	215.1	1157.0	941.9	866.9	75.0	0.3624	14.58_{24}	0.06858_{115}	**246**
247	28.319_{505}	216.1	1157.3	941.2	866.1	75.1	0.3639	14.34_{23}	0.06973_{116}	**247**
248	28.824_{512}	217.1	1157.6	940.5	865.3	75.2	0.3653	14.11_{23}	0.07089_{118}	**248**
249	29.336_{520}	218.1	1157.9	939.8	864.5	75.3	0.3667	13.88_{23}	0.07207_{120}	**249**
250	29.856_{528}	219.1	1158.2	939.1	863.8	75.3	0.3681	13.65_{22}	0.07327_{121}	**250**
251	30.384_{535}	220.1	1158.5	938.4	863.0	75.4	0.3695	13.43_{22}	0.07448_{123}	**251**
252	30.919_{543}	221.1	1158.8	937.7	862.2	75.5	0.3709	13.21_{22}	0.07571_{126}	**252**
253	31.462_{550}	222.1	1159.1	937.0	861.4	75.6	0.3724	12.99_{21}	0.07697_{128}	**253**
254	32.012_{559}	223.1	1159.4	936.3	860.7	75.6	0.3738	12.78_{21}	0.07825_{128}	**254**
255	32.571_{566}	224.1	1159.7	935.6	859.9	75.7	0.3752	12.57_{20}	0.07953_{129}	**255**
256	33.137_{574}	225.1	1160.0	934.9	859.1	75.8	0.3766	12.37_{20}	0.08082_{132}	**256**
257	33.711_{583}	226.2	1160.3	934.1	858.2	75.9	0.3780	12.17_{19}	0.08214_{133}	**257**
258	34.294_{590}	227.2	1160.6	933.4	857.5	75.9	0.3794	11.98_{19}	0.08347_{135}	**258**
259	34.884_{599}	228.2	1160.9	932.7	856.7	76.0	0.3808	11.79_{19}	0.08482_{137}	**259**
260	35.483_{607}	229.2	1161.2	932.0	855.9	76.1	0.3822	11.60_{18}	0.08619_{138}	**260**
261	36.090_{616}	230.2	1161.5	931.3	855.1	76.2	0.3836	11.42_{18}	0.08757_{140}	**261**
262	36.706_{624}	231.2	1161.9	930.7	854.4	76.3	0.3850	11.24_{18}	0.08897_{142}	**262**
263	37.330_{633}	232.2	1162.2	930.0	853.7	76.3	0.3864	11.06_{17}	0.09039_{143}	**263**

SATURATED STEAM—*Continued.*

Temperature, Degrees Fahr. t	Pressure, Pounds per Square Inch. p	Heat of the Liquid. q	Total Heat. λ	Heat of Vaporization. r	Heat equivalent of Internal Work. ρ	Heat equivalent of External Work. Apu	Entropy of the Liquid. $\int\frac{cdt}{T}$	Specific Volume. s	Density. Weight, in Pounds, of a Cubic Foot. γ	Temperature, Degrees Fahr. t
264	37.963_{641}	233.2	1162.5	929.3	852.9	76.4	0.3878	10.89_{17}	0.09182_{145}	**264**
265	38.604_{651}	234.2	1162.8	928.6	852.1	76.5	0.3891	10.72_{17}	0.09327_{147}	**265**
266	39.255_{659}	235.2	1163.1	927.9	851.3	76.6	0.3906	10.55_{16}	0.09474_{150}	**266**
267	39.914_{668}	236.2	1163.4	927.2	850.6	76.6	0.3919	10.39_{16}	0.09624_{151}	**267**
268	40.582_{677}	237.2	1163.7	926.5	849.8	76.7	0.3933	10.23_{16}	0.09775_{152}	**268**
269	41.259_{686}	238.2	1164.0	925.8	849.0	76.8	0.3947	10.07_{15}	0.09927_{153}	**269**
270	41.945_{695}	239.3	1164.3	925.0	848.1	76.9	0.3961	9.918_{152}	0.1008_{16}	**270**
271	42.640_{705}	240.3	1164.6	924.3	847.4	76.9	0.3975	9.766_{149}	0.1024_{16}	**271**
272	43.345_{714}	241.3	1164.9	923.6	846.6	77.0	0.3988	9.617_{146}	0.1040_{16}	**272**
273	44.059_{723}	242.3	1165.2	922.9	845.8	77.1	0.4002	9.471_{143}	0.1056_{16}	**273**
274	44.782_{733}	243.3	1165.5	922.2	845.0	77.2	0.4016	9.328_{141}	0.1072_{16}	**274**
275	45.515_{743}	244.3	1165.8	921.5	844.2	77.3	0.4030	9.187_{138}	0.1088_{17}	**275**
276	46.258_{753}	245.3	1166.1	920.8	843.5	77.3	0.4043	9.049_{136}	0.1105_{17}	**276**
277	47.011_{762}	246.3	1166.4	920.1	842.7	77.4	0.4057	8.913_{133}	0.1122_{17}	**277**
278	47.773_{772}	247.3	1166.7	919.4	841.9	77.5	0.4071	8.780_{131}	0.1139_{17}	**278**
279	48.545_{783}	248.3	1167.0	918.7	841.1	77.6	0.4084	8.649_{128}	0.1156_{17}	**279**
280	49.328_{792}	249.3	1167.3	918.0	840.4	77.6	0.4098	8.521_{126}	0.1173_{18}	**280**
281	50.12_{80}	250.3	1167.6	917.3	839.6	77.7	0.4112	8.395_{124}	0.1191_{18}	**281**
282	50.92_{82}	251.4	1168.0	916.6	838.8	77.8	0.4125	8.271_{122}	0.1209_{18}	**282**
283	51.74_{82}	252.4	1168.3	915.9	838.0	77.9	0.4139	8.149_{119}	0.1227_{18}	**283**
284	52.56_{83}	253.4	1168.6	915.2	837.2	78.0	0.4152	8.030_{117}	0.1245_{19}	**284**
285	53.39_{85}	254.4	1168.9	914.5	836.5	78.0	0.4166	7.913_{116}	0.1264_{19}	**285**
286	54.24_{85}	255.4	1169.2	913.8	835.7	78.1	0.4179	7.797_{113}	0.1283_{19}	**286**
287	55.09_{87}	256.4	1169.5	913.1	834.9	78.2	0.4193	7.684_{111}	0.1302_{19}	**287**
288	55.96_{87}	257.4	1169.8	912.4	834.1	78.3	0.4206	7.573_{109}	0.1321_{19}	**288**
289	56.83_{89}	258.4	1170.1	911.7	833.4	78.3	0.4220	7.464_{108}	0.1340_{19}	**289**
290	57.72_{90}	259.4	1170.4	911.0	832.6	78.4	0.4233	7.356_{105}	0.1359_{20}	**290**
291	58.62_{91}	260.4	1170.7	910.3	831.8	78.5	0.4247	7.251_{103}	0.1379_{20}	**291**
292	59.53_{92}	261.4	1171.0	909.6	831.0	78.6	0.4260	7.148_{102}	0.1399_{20}	**292**
293	60.45_{93}	262.4	1171.3	908.9	830.3	78.6	0.4273	7.046_{100}	0.1419_{21}	**293**
294	61.38_{95}	263.4	1171.6	908.2	829.5	78.7	0.4287	6.946_{99}	0.1440_{21}	**294**
295	62.33_{95}	264.5	1171.9	907.4	828.6	78.8	0.4300	6.847_{97}	0.1461_{21}	**295**
296	63.28_{97}	265.5	1172.2	906.7	827.8	78.9	0.4313	6.750_{95}	0.1482_{21}	**296**
297	64.25_{98}	266.5	1172.5	906.0	827.0	79.0	0.4327	6.655_{93}	0.1503_{21}	**297**
298	65.23_{99}	267.5	1172.8	905.3	826.3	79.0	0.4340	6.562_{92}	0.1524_{21}	**298**
299	66.22_{100}	268.5	1173.1	904.6	825.5	79.1	0.4353	6.470_{90}	0.1545_{22}	**299**
300	67.22_{102}	269.5	1173.4	903.9	824.7	79.2	0.4366	6.380_{88}	0.1567_{22}	**300**
301	68.24_{103}	270.5	1173.7	903.2	823.9	79.3	0.4380	6.292_{87}	0.1589_{22}	**301**
302	69.27_{103}	271.5	1174.1	902.6	823.3	79.3	0.4393	6.205_{86}	0.1611_{23}	**302**
303	70.30_{106}	272.5	1174.4	901.9	822.5	79.4	0.4406	6.119_{84}	0.1634_{23}	**303**

SATURATED STEAM — *Continued.*

Temperature, Degrees Fahr. t	Pressure, Pounds per Square Inch. p	Heat of the Liquid. q	Total Heat. λ	Heat of Vaporization. r	Heat equivalent of Internal Work. ρ	Heat equivalent of External Work. Apu	Entropy of the Liquid. $\int \frac{cdt}{T}$	Specific Volume. s	DENSITY. Weight, in Pounds, of one Cubic Foot. γ	Temperature, Degrees Fahr. t
304	71.36_{106}	273.5	1174.7	901.2	821.7	79.5	0.4419	6.035_{83}	0.1657_{23}	304
305	72.42_{108}	274.5	1175.0	900.5	820.9	79.6	0.4433	5.952_{81}	0.1680_{23}	305
306	73.50_{109}	275.5	1175.3	899.8	820.1	79.7	0.4446	5.871_{80}	0.1703_{24}	306
307	74.59_{110}	276.6	1175.6	899.0	819.3	79.7	0.4459	5.791_{79}	0.1727_{24}	307
308	75.69_{111}	277.6	1175.9	898.3	818.5	79.8	0.4472	5.712_{78}	0.1751_{24}	308
309	76.80_{113}	278.6	1176.2	897.6	817.7	79.9	0.4485	5.634_{76}	0.1775_{24}	309
310	77.93_{114}	279.6	1176.5	896.9	817.0	79.9	0.4498	5.558_{74}	0.1799_{24}	310
311	79.07_{116}	280.6	1176.8	896.2	816.2	80.0	0.4511	5.484_{74}	0.1823_{25}	311
312	80.23_{116}	281.6	1177.1	895.5	815.4	80.1	0.4524	5.410_{73}	0.1848_{25}	312
313	81.39_{118}	282.7	1177.4	894.7	814.5	80.2	0.4538	5.337_{71}	0.1873_{26}	313
314	82.57_{120}	283.7	1177.7	894.0	813.8	80.2	0.4552	5.266_{71}	$0\ 1899_{26}$	314
315	83.77_{121}	284.8	1178.0	893.2	812.9	80.3	0.4565	5.195_{69}	0.1925_{26}	315
316	84.98_{122}	285.8	1178.3	892.5	812.1	80.4	0.4579	5.126_{68}	0.1951_{26}	316
317	86.20_{123}	286.9	1178.6	891.7	811.3	80.4	0.4592	5.058_{67}	0.1977_{27}	317
318	87.43_{125}	287.9	1178.9	891.0	810.5	80.5	0.4606	4.991_{66}	0.2004_{27}	318
319	88.68_{127}	289.0	1179.2	890.2	809.6	80.6	0.4619	4.925_{64}	0.2031_{27}	319
320	89.95_{128}	290.0	1179.5	889.5	808.8	80.7	0.4633	4.861_{64}	0.2058_{27}	320
321	91.23_{129}	291.0	1179.8	888.8	808.1	80.7	0.4646	4.797_{62}	0.2085_{27}	321
322	92.52_{130}	292.1	1180.2	888.1	807.3	80.8	0.4659	4.735_{62}	0.2112_{28}	322
323	93.82_{132}	293.1	1180.5	887.4	806.5	80.9	0.4672	4.673_{61}	0.2140_{28}	323
324	95.14_{134}	294.2	1180.8	886.6	805.7	80.9	0.4686	612_{60}	0.2168_{29}	324
325	96.48_{135}	295.2	1181.1	885.9	804.9	81.0	0.4699	552_{59}	0.2197_{29}	325
326	97.83_{137}	296.3	1181.4	885.1	804.1	81.1	0.4713	4.493_{57}	0.2226_{29}	326
327	99.20_{138}	297.3	1181.7	884.4	803.3	81.1	0.4726	4.436_{57}	0.2255_{29}	327
328	100.58_{139}	298.4	1182.0	883.6	802.4	81.2	0.4739	4.379_{56}	0.2284_{29}	328
329	101.97_{141}	299.4	1182.3	882.9	801.6	81.3	0.4752	4.323_{56}	0.2313_{30}	329
330	103.38_{143}	300.5	1182.6	882.1	800.8	81.3	0.4766	4.267_{54}	0.2343_{31}	330
331	104.81_{144}	301.5	1182.9	881.4	800.0	81.4	0.4779	4.213_{54}	0.2374_{30}	331
332	106.25_{145}	302.6	1183.2	880.6	799.1	81.5	0.4792	4.159_{52}	0.2404_{31}	332
333	107.70_{147}	303.6	1183.5	879.9	798.4	81.5	0.4805	4.107_{52}	0.2435_{31}	333
334	109.17_{149}	304.6	1183.8	879.2	797.6	81.6	0.4818	4.055_{51}	0.2466_{32}	334
335	110.66_{151}	305.7	1184.1	878.4	796.7	81.7	0.4832	4.004_{50}	0.2498_{31}	335
336	112.17_{152}	306.7	1184.4	877.7	796.0	81.7	0.4845	3.954_{50}	0.2529_{32}	336
337	113.69_{153}	307.8	1184.7	876.9	795.1	81.8	0.4858	3.904_{49}	0.2561_{33}	337
338	115.22_{155}	308.8	1185.0	876.2	794.3	81.9	0.4871	3.855_{48}	0.2594_{33}	338
339	116.77_{157}	309.9	1185.3	875.4	793.5	81.9	0.4884	3.807_{47}	0.2627_{33}	339
340	118.34_{159}	310.9	1185.6	874.7	792.7	82.0	0.4897	3.760_{47}	0.2660_{33}	340
341	119.93_{160}	312.0	1185.9	873.9	791.8	82.1	0.4910	3.713_{45}	0.2693_{33}	341
342	121.53_{162}	313.0	1186.3	873.3	791.2	82.1	0.4923	3.668_{45}	0.2726_{34}	342
343	123.15_{163}	314.1	1186.6	872.5	790.3	82.2	0.4936	8.623_{45}	0.2760_{35}	343

SATURATED STEAM—*Continued.*

Temperature, Degrees Fahr. t	Pressure, Pounds per Square Inch. p	Heat of the Liquid. q	Total Heat. λ	Heat of Vaporization. r	Heat equivalent of Internal Work. ρ	Heat equivalent of External Work. Apu	Entropy of the Liquid. $\int\frac{cdt}{T}$	Specific Volume. s	Density. Weight, Pounds, of one Cubic Foot. γ	Temperature, Degrees Fahr. t
344	124.78_{165}	315.1	1186.9	871.8	789.5	82.3	0.4949	3.578_{44}	0.2795_{35}	344
345	126.43_{167}	316.1	1187.2	871.1	788.8	82.3	0.4962	3.534_{43}	0.2830_{35}	345
346	128.10_{169}	317.2	1187.5	870.3	787.9	82.4	0.4975	3.491_{42}	0.2865_{35}	346
347	129.79_{170}	318.2	1187.8	869.6	787.1	82.5	0.4988	3.449_{42}	0.2900_{35}	347
348	131.49_{172}	319.3	1188.1	868.8	786.3	82.5	0.5001	3.407_{42}	0.2935_{36}	348
349	133.21_{174}	320.3	1188.4	868.1	785.5	82.6	0.5014	3.365_{41}	0.2971_{37}	349
350	134.95_{176}	321.4	1188.7	867.3	784.7	82.6	0.5027	3.324_{40}	0.3008_{37}	350
351	136.71_{177}	322.4	1189.0	866.6	783.9	82.7	0.5040	3.284_{39}	0.3045_{37}	351
352	138.48_{179}	323.5	1189.3	865.8	783.0	82.8	0.5053	3.245_{39}	0.3082_{37}	352
353	140.27_{181}	324.5	1189.6	865.1	782.3	82.8	0.5066	3.206_{38}	0.3119_{38}	353
354	142.08_{183}	325.6	1189.9	864.3	781.4	82.9	0.5078	3.168_{38}	0.3157_{38}	354
355	143.91_{184}	326.6	1190.2	863.6	780.7	82.9	0.5091	3.130_{38}	0.3195_{39}	355
356	145.75_{187}	327.7	1190.5	862.8	779.8	83.0	0.5104	3.092_{36}	0.3234_{38}	356
357	147.62_{188}	328.7	1190.8	862.1	779.0	83.1	0.5117	3.056_{36}	0.3272_{39}	357
358	149.50_{190}	329.7	1191.1	861.4	778.3	83.1	0.5130	3.020_{36}	0.3311_{40}	358
359	151.40_{193}	330.8	1191.4	860.6	777.4	83.2	0.5142	2.984_{35}	0.3351_{40}	359
360	153.33_{194}	331.8	1191.7	859.9	776.7	83.2	0.5155	2.949_{35}	0.3391_{40}	360
361	155.27_{195}	332.9	1192.0	859.1	775.8	83.3	0.5168	2.914_{34}	0.3431_{41}	361
362	157.22_{198}	333.9	1192.4	858.5	775.2	83.3	0.5181	2.880_{34}	0.3472_{41}	362
363	159.20_{200}	335.0	1192.7	857.7	774.3	83.4	0.5193	2.846_{33}	0.3513_{42}	363
364	161.20_{202}	336.0	1193.0	857.0	773.5	83.5	0.5206	2.813_{33}	0.3555_{42}	364
365	163.22_{203}	337.1	1193.3	856.2	772.7	83.5	0.5219	2.780_{32}	0.3597_{42}	365
366	165.25_{206}	338.1	1193.6	855.5	771.9	83.6	0.5231	2.748_{32}	0.3639_{43}	366
367	167.31_{208}	339.2	1193.9	854.7	771.1	83.6	0.5244	2.716_{31}	0.3682_{43}	367
368	169.39_{209}	340.2	1194.2	854.0	770.4	83.6	0.5257	2.685_{31}	0.3725_{43}	368
369	171.48_{212}	341.3	1194.5	853.2	769.5	83.7	0.5269	2.654_{31}	0.3768_{44}	369
370	173.60_{214}	342.3	1194.8	852.5	768.7	83.8	0.5282	2.623_{30}	0.3812_{44}	370
371	175.74_{215}	343.3	1195.1	851.8	768.0	83.8	0.5294	2.593_{30}	0.3856_{45}	371
372	177.89_{218}	344.4	1195.4	851.0	767.1	83.9	0.5307	2.563_{29}	0.3901_{45}	372
373	180.07_{220}	345.5	1195.7	850.2	766.3	83.9	0.5320	2.534_{29}	0.3946_{46}	373
374	182.27_{222}	346.5	1196.0	849.5	765.5	84.0	0.5332	2.505_{29}	0.3992_{46}	374
375	184.49_{224}	347.5	1196.3	848.8	764.8	84.0	0.5345	2.476_{28}	0.4038_{46}	375
376	186.73_{226}	348.6	1196.6	848.0	763.9	84.1	0.5357	2.448_{28}	0.4084_{47}	376
377	188.99_{228}	349.6	1196.9	847.3	763.2	84.1	0.5370	2.420_{27}	0.4131_{47}	377
378	191.27_{231}	350.6	1197.2	846.6	762.4	84.2	0.5382	2.393_{27}	0.4178_{49}	378
379	193.58_{233}	351.7	1197.5	845.8	761.6	84.2	0.5395	2.366_{28}	0.4227_{49}	379
380	195.91_{234}	352.8	1197.8	845.0	760.8	84.2	0.5407	2.338_{25}	0.4276_{47}	380
381	198.25_{237}	353.8	1198.1	844.3	760.0	84.3	0.5420	2.313_{26}	0.4323_{49}	381
382	200.62_{239}	354.9	1198.5	843.6	759.3	84.3	0.5432	2.287_{25}	0.4372_{49}	382
383	203.01_{242}	355.9	1198.8	842.9	758.5	84.4	0.5444	2.262_{25}	0.4421_{49}	383

SATURATED STEAM — *Continued.*

Temperature, Degrees Fahr. t	Pressure, Pounds per Square Inch. p	Heat of the Liquid. q	Total Heat. λ	Heat of Vaporization. r	Heat equivalent of Internal Work. ρ	Heat equivalent of External Work. Apu	Entropy of the Liquid. $\int \frac{cdt}{T}$	Specific Volume. s	DENSITY. Weight, in Pounds, of one Cubic Foot. γ	Temperature, Degrees Fahr. t
384	205.43_{244}	356.9	1199.1	842.2	757.8	84.4	0.5457	2.237_{25}	0.4470_{51}	384
385	207.87_{246}	358.0	1199.4	841.4	756.9	84.5	0.5469	2.212_{25}	0.4521_{51}	385
386	210.33_{248}	359.0	1199.7	840.7	756.2	84.5	0.5481	2.187_{24}	0.4572_{51}	386
387	212.81_{250}	360.1	1200.0	839.9	755.3	84.6	0.5494	2.163_{24}	0.4623_{52}	387
388	215.31_{253}	361.1	1200.3	839.2	754.6	84.6	0.5506	2.139_{24}	0.4675_{53}	388
389	217.84_{255}	362.2	1200.6	838.4	753.8	84.6	0.5518	2.115_{23}	0.4728_{52}	389
390	220.39_{257}	363.2	1200.9	837.7	753.0	84.7	0.5531	2.092_{23}	0.4780_{53}	390
391	222.96_{260}	364.3	1201.2	836.9	752.2	84.7	0.5543	2.069_{23}	0.4833_{54}	391
392	225.56_{263}	365.3	1201.5	836.2	751.4	84.8	0.5555	2.046_{22}	0.4887_{54}	392
393	228.19_{264}	366.4	1201.8	835.4	750.6	84.8	0.5568	2.024_{22}	0.4941_{55}	393
394	230.83_{267}	367.4	1202.1	834.7	749.9	84.8	0.5580	2.002_{22}	0.4996_{55}	394
395	233.50_{269}	368.4	1202.4	834.0	749.1	84.9	0.5592	1.980_{22}	0.5051_{56}	395
396	236.19_{272}	369.5	1202.7	833.2	748.3	84.9	0.5604	1.958_{21}	0.5107_{56}	396
397	238.91_{274}	370.5	1203.0	832.5	747.6	84.9	0.5616	1.937_{21}	0.5163_{56}	397
398	241.65_{277}	371.6	1203.3	831.7	746.7	85.0	0.5629	1.916_{21}	0.5219_{58}	398
399	244.42_{279}	372.6	1203.6	831.0	746.0	85.0	0.5641	1.895_{21}	0.5277_{59}	399
400	247.21_{282}	373.7	1203.9	830.2	745.2	85.0	0.5653	1.874_{20}	0.5336_{58}	400
401	250.03_{284}	374.7	1204.2	829.5	744.5	85.0	0.5665	1.854_{20}	0.5394_{58}	401
402	252.87_{287}	375.8	1204.6	828.8	743.7	85.1	0.5677	1.834_{20}	0.5452_{60}	402
403	255.74_{289}	376.8	1204.9	828.1	743.0	85.1	0.5689	1.814_{20}	0.5112_{60}	403
404	258.63_{292}	377.9	1205.2	827.3	742.2	85.1	0.5701	1.794_{19}	0.5572_{61}	404
405	261.55_{295}	378.9	1205.5	826.6	741.4	85.2	0.5714	1.775_{19}	0.5633_{62}	405
406	264.50_{297}	380.0	1205.8	825.8	740.6	85.2	0.5726	1.756_{19}	0.5695_{61}	406
407	267.47_{300}	381.0	1206.1	825.1	739.9	85.2	0.5738	1.737_{18}	0.5756_{62}	407
408	270.47_{302}	382.0	1206.4	824.4	739.2	85.2	0.5741	1.719_{19}	0.5818_{63}	408
409	273.49_{305}	383.1	1206.7	823.6	738.3	85.3	0.5762	1.700_{18}	0.5881_{64}	409
410	276.54_{308}	384.1	1207.0	822.9	737.6	85.3	0.5774	1.682_{18}	0.5945_{65}	410
411	279.62_{311}	385.2	1207.3	822.1	736.8	85.3	0.5786	1.664_{18}	0.601_{6}	411
412	282.73_{313}	386.2	1207.6	821.4	736.1	85.3	0.5798	1.646_{17}	0.607_{7}	412
413	285.86_{316}	387.3	1207.9	820.6	735.3	85.3	0.5810	1.629_{17}	0.614_{6}	413
414	289.02_{319}	388.3	1208.2	819.9	734.5	85.4	0.5822	1.612_{17}	0.620_{7}	414
415	292.21_{321}	389.4	1208.5	819.1	733.7	85.4	0.5834	1.595_{17}	0.627_{7}	415
416	295.42_{325}	390.4	1208.8	818.4	733.0	85.4	0.5846	1.578_{17}	0.634_{7}	416
417	298.67_{327}	391.5	1209.1	817.6	732.2	85.4	0.5858	1.561_{16}	0.641_{6}	417
418	301.94_{330}	392.5	1209.4	816.9	731.5	85.4	0.5870	1.545_{17}	0.647_{7}	418
419	305.24_{333}	393.6	1209.7	816.1	730.7	85.4	0.5881	1.528_{16}	0.654_{7}	419
420	308.57_{336}	394.6	1210.0	815.4	730.0	85.4	0.5893	1.512_{16}	0.661_{7}	420
421	311.93_{338}	395.6	1210.3	814.7	729.3	85.4	0.5905	1.496_{16}	0.668_{8}	421
422	315.31_{342}	396.7	1210.7	814.0	728.5	85.5	0.5917	1.480_{15}	0.676_{7}	422
423	318.73_{345}	397.7	1211.0	813.3	727.8	85.5	0.5929	1.465_{16}	0.683_{7}	423

SATURATED STEAM — *Continued.*

Temperature, Degrees Fahr. t	Pressure, Pounds per Square Inch. p	Diff.	Heat of the Liquid. q	Total Heat. λ	Heat of Vaporization. r	Heat equivalent of Internal Work. ρ	Heat equivalent of External Work. Apu	Entropy of the Liquid. $\int\frac{cdt}{T}$	Specific Volume. s	Diff.	Density. Weight, in Pounds, of one Cubic Foot. γ	Diff.	Temperature, Degrees Fahr. t
424	322.18	347	398.8	1211.3	812.5	727.0	85.5	0.5941	1.449	15	0.690	7	**424**
425	325.65	351	399.8	1211.6	811.8	726.3	85.5	0.5953	1.434	15	0.697	8	**425**
426	329.16	354	400.9	1211.9	811.0	725.5	85.5	0.5964	1.419	15	0.705	7	**426**
427	332.70	356	401.9	1212.2	810.3	724.8	85.5	0.5976	1.404	14	0.712	7	**427**
428	336.26		403.0	1212.5	809.5	724.0	85.5	0.5988	1.390		0.719		**428**

TABLE II.

SATURATED STEAM.

ENGLISH UNITS.

Pressure, Pounds per Square Inch. p	Temperature, Degrees Fahr. t	Heat of the Liquid. q	Total Heat. λ	Heat of Vaporization. r	Heat-equivalent of Internal Work. ρ	Heat-equivalent of External Work. Apu	Entropy of Liquid. $\int \frac{cdt}{T}$	Specific Volume. s	DENSITY. Weight, in Pounds, of one Cubic Foot. γ	Pressure, Pounds per Square Inch. p
1	101.99_{2428}	70.0	1113.1	1043.0	981.1	61.9	0.1329	$334.6_{161.0}$	0.00299_{277}	1
2	126.27_{1535}	94.4	1120.5	1026.1	961.9	64.2	0.1754	$173.6_{55.2}$	0.00576_{268}	2
3	141.62_{1147}	109.8	1125.1	1015.3	949.5	65.8	0.2013	$118.4_{28.1}$	0.00844_{263}	3
4	153.09_{925}	121.4	1128.6	1007.2	940.4	6.	0.2203	$90.31_{17.09}$	0.01107_{259}	
5	162.34_{780}	130.7	1131.5	1000.8	933.1	7.8	0.2353	$73.22_{11.55}$	0.01366_{256}	
6	170.14_{676}	138.6	1133.8	995.2	926.7	68.5	0.2480	$61.67_{8.30}$	0.01622_{252}	[illegible]
7	176.90_{602}	145.4	1135.9	990.5	921.4	1	0.2587	$53.37_{6.30}$	0.01874_{251}	7
8	182.92_{541}	151.5	1137.7	986.2	916.5	7	0.2682	$47.07_{4.94}$	0.02125_{249}	8
9	188.33_{492}	156.9	1139.4	982.5	912.4	69.1	0.2766	$42.13_{3.97}$	0.02374_{247}	9
10	193.25_{453}	161.9	1140.9	979.0	908.4	70.6	0.2842	$38.16_{3.28}$	0.02621_{245}	10
11	197.78_{420}	166.5	1142.3	975.8	904.8	71.	0.2912	$34.88_{2.74}$	0.02866_{245}	11
12	201.98_{391}	170.7	1143.6	972.9	901.5	71.0	0.2976	$32.14_{2.32}$	0.03111_{244}	12
13	205.89_{368}	174.6	1144.7	970.1	898.4	71.4	0.3035	$29.82_{2.03}$	0.03355_{245}	13
14	209.57_{346}	178.3	1145.8	967.5	895.5	72.0	0.3091	$27.79_{1.64}$	0.03600_{226}	14
15	213.03_{329}	181.8	1146.9	965.1	892.6	72.5	0.3143	$26.15_{1.56}$	0.03826_{241}	15
16	216.32_{312}	185.1	1147.9	962.8	890.0	72.8	0.3192	$24.59_{1.37}$	0.04067_{240}	16
17	219.44_{296}	188.3	1148.9	960.6	887.6	73.0	0.3238	$23.22_{1.22}$	0.04307_{240}	17
18	222.40_{284}	191.3	1149.8	958.5	885.3	73.2	0.3282	$22.00_{1.10}$	0.04547_{239}	18
19	225.24_{271}	194.1	1150.7	956.6	883.2	73.4	0.3324	$20.90_{0.99}$	0.04786_{237}	19
20	227.95_{260}	196.9	1151.5	954.6	881.0	73.6	0.3363	19.91_{90}	0.05023_{236}	20
21	230.55_{251}	199.5	1152.3	952.8	879.0	73.8	0.3401	19.01_{81}	0.05259_{236}	21
22	233.06_{241}	202.0	1153.0	951.0	877.0	73.0	0.3438	18.20_{75}	0.05495_{236}	22
23	235.47_{232}	204.5	1153.7	949.2	875.0	74.2	0.3473	17.45_{69}	0.05731_{235}	23
24	237.79_{225}	206.8	1154.4	947.6	873.2	74.	0.3506	16.76_{3}	0.05966_{233}	24
25	240.04_{217}	209.1	1155.1	946.0	871.5	74.4	0.3539	16.13_{8}	0.06199_{233}	25
26	242.21_{211}	211.2	1155.8	944.6	869.9	74.5	0.3570	15.55_{65}	0.06432_{234}	26
27	244.32_{204}	213.4	1156.5	943.1	868.2	74.9	0.3600	15.00_{51}	0.06666_{233}	27
28	246.36_{198}	215.4	1157.1	941.7	866.7	75.0	0.3629	14.49_{46}	0.06899_{231}	28
29	248.34_{193}	217.4	1157.7	940.3	865.1	75.2	0.3657	14.03_{44}	0.07130_{230}	29
30	250.27_{188}	219.4	1158.3	938.9	863.6	75.3	0.3685	13.59_{41}	0.07360_{230}	0
31	252.15_{183}	221.3	1158.8	937.5	862.0	75.5	0.3712	13.18_{40}	0.07590_{231}	31
32	253.98_{178}	223.1	1159.4	936.3	860.7	75.6	0.3737	12.78_{37}	0.07821_{230}	32
33	255.76_{174}	224.9	1159.9	935.0	859.2	75.8	0.3762	12.41_{34}	0.08051_{229}	33

SATURATED STEAM — *Continued.*

Pressure, Pounds per Square Inch. p	Temperature, Degrees Fahr. t	Heat of the Liquid. q	Total Heat. λ	Heat of Vaporization. r	Heat equivalent of Internal Work. ρ	Heat equivalent of External Work. Apu	Entropy of the Liquid. $\int\frac{cdt}{T}$	Specific Volume. s	Density. Weight, in Pounds, of one Cubic Foot. γ	Pressure, Pounds per Square Inch. p
34	257.50_{169}	226.7	1160.4	933.7	857.8	75.9	0.3787	12.07_{32}	0.08280_{228}	**34**
35	259.19_{166}	228.4	1161.0	932.6	856.6	76.0	0.3811	11.75_{30}	0.08508_{228}	**35**
36	260.85_{162}	230.0	1161.5	931.5	855.3	76.2	0.3834	11.45_{29}	0.08736_{228}	**36**
37	262.47_{159}	231.7	1162.0	930.3	854.0	76.3	0.3856	11.16_{28}	0.08964_{227}	**37**
38	264.06_{155}	233.3	1162.5	929.2	852.8	76.4	0.3878	10.88_{26}	0.09191_{226}	**38**
39	265.61_{151}	234.8	1163.0	928.2	851.7	76.5	0.3900	10.62_{25}	0.09417_{227}	**39**
40	267.13_{149}	236.4	1163.4	927.0	850.3	76.7	0.3921	10.37_{24}	0.09644_{225}	**40**
41	268.62_{146}	237.9	1163.9	926.0	849.2	76.8	0.3942	10.13_{22}	0.09869_{221}	**41**
42	270.08_{143}	239.3	1164.3	925.0	848.1	76.9	0.3962	9.906_{216}	0.1009_{23}	**42**
43	271.51_{140}	240.8	1164.8	924.0	847.0	77.0	0.3982	9.690_{206}	0.1032_{22}	**43**
44	272.91_{138}	242.2	1165.2	923.0	845.9	77.1	0.4001	9.484_{197}	0.1054_{23}	**44**
45	274.29_{136}	243.6	1165.6	922.0	844.8	77.2	0.4020	9.287_{190}	0.1077_{22}	**45**
46	275.65_{134}	245.0	1166.0	921.0	843.7	77.3	0.4038	9.097_{183}	0.1099_{23}	**46**
47	276.99_{131}	246.3	1166.4	920.1	842.7	77.4	0.4056	8.914_{174}	0.1122_{22}	**47**
48	278.30_{128}	247.6	1166.8	919.2	841.7	77.5	0.4074	8.740_{167}	0.1144_{22}	**48**
49	279.58_{127}	248.9	1167.2	918.3	840.7	77.6	0.4092	8.573_{159}	0.1166_{22}	**49**
50	280.85_{125}	250.2	1167.6	917.4	839.7	77.7	0.4109	8.414_{155}	0.1188_{23}	**50**
51	282.10_{122}	251.5	1168.0	916.5	838.7	77.8	0.4126	8.259_{149}	0.1211_{22}	**51**
52	283.32_{121}	252.7	1168.4	915.7	837.8	77.9	0.4143	8.110_{142}	0.1233_{22}	**52**
53	284.53_{119}	253.9	1168.7	914.8	836.8	78.0	0.4160	7.968_{138}	0.1255_{22}	**53**
54	285.72_{117}	255.1	1169.1	914.0	835.9	78.1	0.4175	7.829_{133}	0.1277_{22}	**54**
55	286.89_{116}	256.3	1169.4	913.1	834.9	78.2	0.4191	7.696_{128}	0.1299_{22}	**55**
56	288.05_{114}	257.5	1169.8	912.3	834.0	78.3	0.4207	7.568_{125}	0.1321_{23}	**56**
57	289.19_{112}	258.6	1170.1	911.5	833.1	78.4	0.4222	7.443_{120}	0.1344_{22}	**57**
58	290.31_{111}	259.7	1170.5	910.8	832.4	78.4	0.4237	7.323_{115}	0.1366_{21}	**58**
59	291.42_{109}	260.8	1170.8	910.0	831.5	78.5	0.4252	7.208_{112}	0.1387_{22}	**59**
60	292.51_{108}	261.9	1171.2	909.3	830.7	78.6	0.4267	7.096_{109}	0.1409_{22}	**60**
61	293.59_{106}	263.0	1171.5	908.5	829.8	78.7	0.4281	6.987_{105}	0.1431_{22}	**61**
62	294.65_{105}	264.1	1171.8	907.7	828.9	78.8	0.4295	6.882_{103}	0.1453_{22}	**62**
63	295.70_{104}	265.2	1172.1	906.9	828.0	78.9	0.4309	6.779_{99}	0.1475_{22}	**63**
64	296.74_{103}	266.2	1172.4	906.2	827.3	78.9	0.4323	6.680_{97}	0.1497_{22}	**64**
65	297.77_{101}	267.2	1172.7	905.5	826.5	79.0	0.4337	6.583_{93}	0.1519_{22}	**65**
66	298.78_{99}	268.3	1173.0	904.7	825.6	79.1	0.4350	6.490_{89}	0.1541_{21}	**66**
67	299.77_{99}	269.3	1173.3	904.0	824.8	79.2	0.4363	6.401_{87}	0.1562_{22}	**67**
68	300.76_{98}	270.3	1173.6	903.3	824.1	79.2	0.4376	6.314_{86}	0.1584_{22}	**68**
69	301.74_{97}	271.2	1173.9	902.7	823.4	79.3	0.4389	6.228_{84}	0.1606_{22}	**69**
70	302.71_{95}	272.2	1174.3	902.1	822.7	79.4	0.4402	6.144_{81}	0.1628_{21}	**70**
71	303.66_{95}	273.2	1174.6	901.4	821.9	79.5	0.4415	6.063_{79}	0.1649_{22}	**71**
72	304.61_{93}	274.1	1174.9	900.8	821.3	79.5	0.4428	5.984_{76}	0.1671_{22}	**72**
73	305.54_{92}	275.1	1175.2	900.1	820.5	79.6	0.4440	5.908_{74}	0.1693_{21}	**73**

SATURATED STEAM—*Continued.*

Pressure, Pounds per Square Inch. p	Temperature, Degrees Fahr. t	Heat of the Liquid. q	Total Heat. λ	Heat of Vaporization. r	Heat equivalent of Internal Work. ρ	Heat equivalent of External Work. Apu	Entropy of the Liquid. $\int\frac{cdt}{T}$	Specific Volume. s	Density. Weight, Pounds, of one Cubic Foot. γ	Pressure, Pounds per Square Inch. p
74	306.46$_{92}$	276.0	1175.4	899.4	819.7	79.7	0.4452	5.834$_{72}$	0.1714$_{22}$	74
75	307.38$_{90}$	276.9	1175.7	898.8	819.1	79.7	0.4464	5.762$_{71}$	0.1736$_{21}$	75
76	308.28$_{90}$	277.8	1176.0	898 2	818.4	79.8	0.4476	5.691$_{70}$	0.1757$_{22}$	76
77	309.18$_{88}$	278.7	1176.2	897.5	817.6	79.9	0.4487	5.621$_{67}$	0.1779$_{22}$	77
78	310.06$_{88}$	279.6	1176.5	896.9	817.0	79.9	0.4499	5.554$_{66}$	0.1801$_{21}$	78
79	310.94$_{86}$	280.5	1176.8	896.3	816.3	80.0	0.4511	5.488$_{63}$	0.1822$_{21}$	79
80	311.80$_{86}$	281.4	1177.0	895.6	815.5	80.1	0.4522	5.425$_{63}$	0.1843$_{22}$	80
81	312.66$_{85}$	282.3	1177.3	895.0	814.9	80.1	0.4534	5.362$_{61}$	0.1865$_{21}$	81
82	313.51$_{85}$	283.2	1177.6	894.4	814.2	80.2	0.4545	5.301$_{61}$	0.1886$_{22}$	82
83	314.36$_{83}$	284.1	1177.8	893.7	813.4	80.3	0.4557	5.240$_{58}$	0.1908$_{22}$	83
84	315.19$_{83}$	285.0	1178.1	893.1	812.8	80.3	0.4568	5.182$_{57}$	0.1930$_{21}$	84
85	316.02$_{82}$	285.8	1178.3	892.5	812.1	80.4	0.4579	5.125$_{56}$	0.1951$_{22}$	85
86	316.84$_{81}$	286.7	1178.6	891.9	811.5	80.4	0.4590	5.069$_{55}$	0.1973$_{21}$	86
87	317.65$_{80}$	287.5	1178.8	891.3	810.8	80.5	0.4601	5.014$_{53}$	0.1994$_{22}$	87
88	318.45$_{80}$	288.4	1179.1	890.7	810.2	80.5	0.4612	4.961$_{52}$	0.2016$_{21}$	88
89	319.25$_{79}$	289.2	1179.3	890.1	809.5	80.6	0.4622	4.909$_{51}$	0.2037$_{21}$	89
90	320.04$_{77}$	290.0	1179.6	889.6	808.9	80.7	0.4633	4.858$_{50}$	0.2058$_{22}$	90
91	320.83$_{77}$	290.8	1179.8	889.0	808.3	80.7	0.4643	4.808$_{48}$	0.2080$_{21}$	91
92	321.60$_{77}$	291.6	1180.0	888.4	807.6	80.8	0.4653	4.760$_{48}$	0.2101$_{21}$	92
93	322.37$_{77}$	292.4	1180.3	887.9	807.1	80.8	0.4663	4.712$_{47}$	0.2122$_{22}$	93
94	323.14$_{75}$	293.2	1180.5	887.3	806.4	80.9	0.4673	4.665$_{46}$	0.2144$_{21}$	94
95	323.89$_{75}$	294.0	1180.7	886.7	805.8	80.9	0.4683	4.619$_{45}$	0.2165$_{21}$	95
96	324.64$_{74}$	294.8	1181.0	886.2	805.2	81.0	0.4693	4.574$_{44}$	0.2186$_{22}$	96
97	325.38$_{74}$	295.6	1181.2	885.6	804.6	81.0	0.4703	4.530$_{44}$	0 2208$_{21}$	97
98	326.12$_{74}$	296.4	1181.4	885.0	803.9	81.1	0.4713	4.486$_{42}$	0.2229$_{21}$	98
99	326.86$_{72}$	297.1	1181.6	884.5	803.4	81.1	0.4723	4.444$_{41}$	0.2250$_{21}$	99
100	327.58$_{72}$	297.9	1181.9	884.0	802.8	81.2	0.4733	4.403$_{41}$	0.2271$_{22}$	100
101	328.30$_{72}$	298.6	1182.1	883.5	802.3	81.2	0.4743	4.362$_{40}$	0.2293$_{21}$	101
102	329.02$_{71}$	299.4	1182.3	882.9	801.6	81.3	0.4753	4.322$_{40}$	0.2314$_{21}$	102
103	329.73$_{70}$	300.1	1182.5	882.4	801.1	81.3	0.4762	4.282$_{38}$	0.2335$_{21}$	103
104	330.43$_{70}$	300.9	1182.7	881.8	800.4	81.4	0.4771	4.244$_{38}$	0.2356$_{22}$	104
105	331.13$_{70}$	301.6	1182.9	881.3	799.9	81.4	0.4780	4.206$_{37}$	0.2378$_{21}$	105
106	331.83$_{69}$	302.3	1183.1	880.8	799.3	81.5	0.4790	4.169$_{37}$	0.2399$_{21}$	106
107	332.52$_{68}$	302.0	1183.4	880.4	798.9	81.5	0.4799	4.132$_{36}$	0.2420$_{21}$	107
108	333.20$_{68}$	303.8	1183.6	879.8	798.2	81.6	0.4808	4.096$_{35}$	0.2441$_{21}$	108
109	333.88$_{68}$	304.5	1183.8	879.3	797.7	81.6	0.4817	4.061$_{35}$	0.2462$_{22}$	109
110	334.56$_{67}$	305.2	1184.0	878.8	797.1	81.7	0.4826	4.026$_{34}$	0.2484$_{21}$	110
111	335.23$_{66}$	305.9	1184.2	878.3	796.6	81.7	0.4835	3.992$_{33}$	0.2505$_{21}$	111
112	335.89$_{66}$	306.6	1184.4	877.8	796.1	81.7	0.4843	3.959$_{33}$	0.2526$_{21}$	112
113	336.55$_{65}$	307.3	1184.6	877.3	795.5	81.8	0.4852	3.926$_{32}$	0.2547$_{21}$	113

SATURATED STEAM — *Continued.*

Pressure, Pounds per Square Inch. p	Temperature, Degrees Fahr. t	Heat of the Liquid. q	Total Heat. λ	Heat of Vaporization. r	Heat equivalent of Internal Work. ρ	Heat equivalent of External Work. Apu	Entropy of the Liquid. $\int\frac{cdt}{T}$	Specific Volume. s	DENSITY. Weight, in Pounds, of one Cubic Foot. γ	Pressure, Pounds per Square Inch. p
114	337.20_{66}	308.0	1184.8	876.8	795.0	81.8	0.4860	3.894_{32}	0.2568_{21}	**114**
115	337.86_{64}	308.7	1185.0	876.3	794.4	81.9	0.4869	3.862_{31}	0.2589_{21}	**115**
116	338.50_{64}	309.4	1185.2	875.8	793.9	81.9	0.4877	3.831_{30}	0.2610_{21}	**116**
117	339.14_{64}	310.0	1185.4	875.4	793.5	81.9	0.4886	3.801_{31}	0.2631_{22}	**117**
118	339.78_{64}	310.7	1185.6	874.9	792.9	82.0	0.4894	3.770_{30}	0.2653_{21}	**118**
119	340.42_{63}	311.4	1185.8	874.4	792.4	82.0	0.4903	3.740_{29}	0.2674_{21}	**119**
120	341.05_{62}	312.0	1186.0	874.0	791.9	82.1	0.4911	3.711_{28}	0.2695_{20}	**120**
121	341.67_{62}	312.7	1186.2	873.5	791.4	82.1	0.4919	3.683_{28}	0.2715_{21}	**121**
122	342.29_{62}	313.3	1186.3	873.0	790.8	82.2	0.4927	3.655_{28}	0.2736_{21}	**122**
123	342.91_{61}	314.0	1186.5	872.5	790.3	82.2	0.4935	3.627_{28}	0.2757_{22}	**123**
124	343.52_{61}	314.6	1186.7	872.1	789.9	82.2	0.4943	3.599_{27}	0.2779_{21}	**124**
125	344.13_{60}	315.2	1186.9	871.7	789.4	82.3	0.4951	3.572_{26}	0.2800_{20}	**125**
126	344.73_{60}	315.9	1187.1	871.2	788.9	82.3	0.4959	3.546_{26}	0.2820_{21}	**126**
127	345.33_{60}	316.5	1187.3	870.8	788.4	82.4	0.4967	3.520_{26}	0.2841_{21}	**127**
128	345.93_{60}	317.1	1187.4	870.3	787.9	82.4	0.4974	3.494_{25}	0.2862_{21}	**128**
129	346.53_{59}	318.7	1187.6	869.9	787.5	82.4	0.4982	3.469_{25}	0.2883_{21}	**129**
130	347.12_{59}	318.4	1187.8	869.4	786.9	82.5	0.4990	3.444_{25}	0.2904_{21}	**130**
131	347.71_{58}	319.0	1188.0	869.0	786.5	82.5	0.4997	3.419_{24}	0.2925_{21}	**131**
132	348.29_{58}	319.6	1188.2	868.6	786.1	82.5	0.5005	3.395_{24}	0.2946_{21}	**132**
133	348.87_{58}	320.2	1188.4	868.2	785.6	82.6	0.5012	3.371_{24}	0.2967_{21}	**133**
134	349.45_{58}	320.8	1188.5	867.7	785.1	82.6	0.5020	3.347_{24}	0.2988_{21}	**134**
135	350.03_{57}	321.4	1188.7	867.3	784.7	82.6	0.5027	3.323_{23}	0.3009_{21}	**135**
136	350.60_{57}	322.0	1188.9	866.9	784.2	82.7	0.5035	3.300_{23}	0.3030_{21}	**136**
137	351.17_{56}	322.6	1189.0	866.4	783.7	82.7	0.5042	3.277_{22}	0.3051_{21}	**137**
138	351.73_{56}	323.2	1189.2	866.0	783.3	82.7	0.5049	3.255_{21}	0.3072_{20}	**138**
139	352.29_{56}	323.8	1189.4	865.6	782.8	82.8	0.5056	3.234_{22}	0.3092_{21}	**139**
140	352.85_{55}	324.4	1189.5	865.1	782.3	82.8	0.5064	3.212_{21}	0.3113_{21}	**140**
141	353.40_{55}	325.0	1189.7	864.7	781.9	82.8	0.5071	3.191_{21}	0.3134_{21}	**141**
142	353.95_{55}	325.6	1189.9	864.3	781.4	82.9	0.5078	3.170_{21}	0.3155_{21}	**142**
143	354.50_{55}	326.1	1190.1	864.0	781.1	82.9	0.5085	3.149_{21}	0.3176_{21}	**143**
144	355.05_{54}	326.7	1190.2	863.5	780.6	82.9	0.5092	3.128_{21}	0.3197_{21}	**144**
145	355.59_{54}	327.2	1190.4	863.2	780.2	83.0	0.5099	3.107_{20}	0.3218_{21}	**145**
146	356.13_{54}	327.8	1190.6	862.8	779.8	83.0	0.5106	3.087_{19}	0.3239_{20}	**146**
147	356.67_{53}	328.3	1190.7	862.4	779.4	83.0	0.5113	3.068_{19}	0.3259_{21}	**147**
148	357.20_{53}	328.9	1190.9	862.0	778.9	83.1	0.5119	3.049_{19}	0.3280_{20}	**148**
149	357.73_{53}	329.4	1191.0	861.6	778.5	83.1	0.5126	3.030_{19}	0.3300_{21}	**149**
150	358.26_{52}	330.0	1191.2	861.2	778.1	83.1	0.5133	3.011_{19}	0.3321_{21}	**150**
151	358.78_{52}	330.5	1191.4	860.9	777.7	83.2	0.5140	2.992_{19}	0.3342_{21}	**151**
152	359.30_{52}	331.1	1191.5	860.4	777.2	83.2	0.5146	2.973_{18}	0.3363_{21}	**152**
153	359.82_{52}	331.6	1191.7	860.1	776.9	83.2	0.5153	2.955_{18}	0.3384_{21}	**153**

Pressure, Pounds per Square Inch. p	Temperature, Degrees Fahr. t	Heat of the Liquid. q	Total Heat. λ	Heat of Vaporization. r	Heat equivalent of Internal Work. ρ	Heat equivalent of External Work. Apu	Entropy of the Liquid. $\int \frac{cdt}{T}$	Specific Volume. s	Density. Weight, in Pounds, of one Cubic Foot. γ	Pressure, Pounds per Square Inch. p
154	360.34$_{52}$	332.2	1191.8	859.6	776.3	83.3	0.5160	2.937$_{18}$	0.3405$_{21}$	154
155	360.86$_{51}$	332.7	1192.0	859.3	776.0	83.3	0.5166	2.919$_{18}$	0.3426$_{21}$	155
156	361.37$_{51}$	333.3	1192.2	858.9	775.6	83.3	0.5173	2.901$_{17}$	0.3447$_{20}$	156
157	361.88$_{51}$	333.8	1192.3	858.5	775.2	83.3	0.5179	2.884$_{17}$	0.3467$_{21}$	157
158	362.39$_{51}$	334.3	1192.5	858.2	774.8	83.4	0.5186	2.867$_{17}$	0.3488$_{21}$	158
159	362.90$_{50}$	334.9	1192.7	857.8	774.4	83.4	0.5192	2.850$_{17}$	0.3509$_{21}$	159
160	363.40$_{50}$	335.4	1192.8	857.4	774.0	83.4	0.5198	2.833$_{17}$	0.3530$_{21}$	160
161	363.90$_{50}$	335.9	1193.0	857.1	773.7	83.4	0.5205	2.816$_{17}$	0.3551$_{21}$	161
162	364.40$_{50}$	336.4	1193.1	856.7	772.2	83.5	0.5211	2.799$_{16}$	0.3572$_{21}$	162
163	364.90$_{49}$	337.0	1193.3	856.3	772.8	83.5	0.5217	2.783$_{16}$	0.3593$_{21}$	163
164	365.39$_{49}$	337.5	1193.4	855.9	772.4	83.5	0.5224	2.767$_{16}$	0.3614$_{21}$	164
165	365.88$_{49}$	338.0	1193.6	855.6	772.0	83.6	0.5230	2.751$_{15}$	0.3635$_{20}$	165
166	366.37$_{48}$	338.5	1193.7	855.2	771.6	83.6	0.5236	2.736$_{15}$	0.3655$_{20}$	166
167	366.85$_{48}$	339.0	1193.9	854.9	771.3	83.	0.5242	2.721$_{15}$	0.3675$_{20}$	167
168	367.33$_{48}$	339.5	1194.0	854.5	770.9	83.6	0.5248	2.706$_{15}$	0.3695$_{21}$	168
169	367.81$_{48}$	340.0	1194.2	854.2	770.5	83.7	0.5254	2.691$_{15}$	0.3716$_{21}$	169
170	368.29$_{48}$	340.5	1194.3	853.8	770.1	83.7	0.5260	2.676$_{15}$	0.3737$_{21}$	170
171	368.77$_{47}$	341.0	1194.4	853.4	769.7	83.7	0.5266	2.661$_{14}$	0.3758$_{20}$	171
172	369.24$_{47}$	341.5	1194.6	853.1	769.4	83.7	0.5272	2.647$_{15}$	0.3778$_{21}$	172
173	369.71$_{47}$	342.0	1194.7	852.7	768.9	83.8	0.5278	2.632$_{14}$	0.3799$_{21}$	173
174	370.18$_{47}$	342.5	1194.8	852.3	768.5	83.8	0.5284	2.618$_{15}$	0.3820$_{21}$	174
175	370.65$_{47}$	343.0	1195.0	852.0	768.2	83.8	0.5290	2.603$_{14}$	0.3841$_{21}$	175
176	371.12$_{47}$	343.5	1195.1	851.6	767.8	83.8	0.5296	2.589$_{14}$	0.3862$_{21}$	176
177	371.59$_{46}$	344.0	1195.3	851.3	767.5	83.8	0.5302	2.575$_{14}$	0.3883$_{21}$	177
178	372.05$_{46}$	344.4	1195.4	851.0	767.1	83.9	0.5308	2.561$_{13}$	0.3904$_{21}$	178
179	372.51$_{46}$	344.9	1195.6	850.7	766.8	83.9	0.5313	2.548$_{13}$	0.3925$_{20}$	179
180	372.97$_{46}$	345.4	1195.7	850.3	766.4	83.9	0.5319	2.535$_{13}$	0.3945$_{21}$	180
181	373.43$_{45}$	345.9	1195.9	850.0	766.1	83.9	0.5325	2.522$_{14}$	0.3966$_{21}$	181
182	373.88$_{45}$	346.4	1196.0	849.6	765.6	84.0	0.5331	2.508$_{13}$	0.3987$_{21}$	182
183	374.33$_{45}$	346.8	1196.1	849.3	765.3	84.0	0.5336	2.495$_{13}$	0.4008$_{21}$	183
184	374.78$_{45}$	347.3	1196.2	848.9	764.9	84.0	0.5342	2.482$_{12}$	0.4029$_{20}$	184
185	375.23$_{45}$	347.8	1196.4	848.6	764.6	84.0	0.5347	2.470$_{13}$	0.4049$_{21}$	185
186	375.68$_{44}$	348.2	1196.5	848.3	764.3	84.0	0.5353	2.457$_{12}$	0.4070$_{20}$	186
187	376.12$_{44}$	348.7	1196.6	847.9	763.8	84.1	0.5359	2.445$_{13}$	0.4090$_{21}$	187
188	376 56$_{44}$	349.2	1196.8	847.6	763.5	84.1	0.5364	2.432$_{12}$	0.4111$_{21}$	188
189	377.00$_{44}$	349.6	1196.9	847.3	763.2	84.1	0.5370	2.420$_{12}$	0.4132$_{21}$	189
190	377.44$_{44}$	350.1	1197.1	847.0	762.9	84.1	0.5375	2.408$_{12}$	0.4153$_{21}$	190
191	377.88$_{44}$	350.5	1197.2	846.7	762.5	84.2	0.5381	2.396$_{11}$	0.4174$_{20}$	191
192	378.32$_{43}$	351.0	1197.3	846.3	762 1	84.2	0.5386	2.385$_{12}$	0.4194$_{21}$	192
193	378.75$_{43}$	351.4	1197.4	846.0	761.8	84.2	0.5391	2.373$_{12}$	0.4215$_{21}$	193

SATURATED STEAM—*Continued.*

Pressure, Pounds per Square Inch. p	Temperature, Degrees Fahr. t	Heat of the Liquid. q	Total Heat. λ	Heat of Vaporization. r	Heat equivalent of Internal Work. ρ	Heat equivalent of External Work. Apu	Entropy of the Liquid. $\int\frac{cdt}{T}$	Specific Volume. s	Density. Weight in Pounds, of One Cubic Foot. γ	Pressure, Pounds per Square Inch. p
194	379.18 $_{43}$	351.9	1197.6	845.7	761.5	84.2	0.5397	2.361 $_{12}$	0.4236 $_{21}$	**194**
195	379.61 $_{43}$	352.4	1197.7	845.3	761.1	84.2	0.5402	2.349 $_{12}$	0.4257 $_{21}$	**195**
196	380.04 $_{43}$	352.8	1197.8	845.0	760.8	84.2	0.5408	2.337 $_{12}$	0.4278 $_{20}$	**196**
197	380.47 $_{42}$	353.3	1198.0	844.7	760.4	84.3	0.5413	2.325 $_{11}$	0.4298 $_{20}$	**197**
198	380.89 $_{42}$	353.7	1198.1	844.4	760.1	84.3	0.5418	2.314 $_{10}$	0.4318 $_{20}$	**198**
199	381.31 $_{42}$	354.1	1198.2	844.1	759.8	84.3	0.5423	2.304 $_{10}$	0.4338 $_{21}$	**199**
200	381.73 $_{42}$	354.6	1198.4	843.8	759.5	84.3	0.5429	2.294 $_{10}$	0.4359 $_{20}$	**200**
201	382.15 $_{42}$	355.0	1198.5	843.5	759.1	84.4	0.5434	2.284 $_{10}$	0.4379 $_{20}$	**201**
202	382.57 $_{42}$	355.4	1198.6	843.2	758.8	84.4	0.5439	2.274 $_{11}$	0.4399 $_{21}$	**202**
203	382.99 $_{42}$	355.9	1198.8	842.9	758.5	84.4	0.5444	2.263 $_{11}$	0.4420 $_{21}$	**203**
204	383.41 $_{41}$	356.3	1198.9	842.6	758.2	84.4	0.5449	2.252 $_{11}$	0.4441 $_{20}$	**204**
205	383.82 $_{41}$	356.8	1199.0	842.2	757.8	84.4	0.5454	2.241 $_{10}$	0.4461 $_{21}$	**205**
206	384.23 $_{41}$	357.2	1199.1	841.9	757.4	84.5	0.5459	2.231 $_{10}$	0.4482 $_{21}$	**206**
207	384.64 $_{41}$	357.6	1199.3	841.7	757.2	84.5	0.5465	2.221 $_{10}$	0.4503 $_{21}$	**207**
208	385.05 $_{41}$	358.0	1199.4	841.4	756.9	84.5	0.5470	2.211 $_{11}$	0.4524 $_{20}$	**208**
209	385.46 $_{41}$	358.5	1199.5	841.0	756.5	84.5	0.5475	2.200 $_{10}$	0.4544 $_{21}$	**209**
210	385.87 $_{41}$	358.9	1199.6	840.7	756.2	84.5	0.5480	2.190 $_{10}$	0.4565 $_{21}$	**210**
211	386.28 $_{40}$	359.3	1199.8	840.5	756.0	84.5	0.5485	2.180 $_{9}$	0.4586 $_{21}$	**211**
212	386.68 $_{40}$	359.7	1199.9	840.2	755.6	84.6	0.5489	2.171 $_{9}$	0.4607 $_{20}$	**212**
213	387.08 $_{40}$	360.1	1200.0	839.9	755.3	84.6	0.5494	2.162 $_{10}$	0.4627 $_{21}$	**213**
214	387.48 $_{40}$	360.6	1200.1	839.5	754.9	84.6	0.5499	2.152 $_{10}$	0.4648 $_{21}$	**214**
215	387.88 $_{40}$	361.0	1200.2	839.2	754.6	84.6	0.5504	2.142 $_{10}$	0.4669 $_{21}$	**215**
216	388.28 $_{39}$	361.4	1200.4	839.0	754.4	84.6	0.5509	2.132 $_{9}$	0.4690 $_{21}$	**216**
217	388.67 $_{39}$	361.8	1200.5	838.7	754.1	84.6	0.5514	2.123 $_{9}$	0.4711 $_{20}$	**217**
218	389.06 $_{39}$	362.2	1200.6	838.4	753.8	84.6	0.5519	2.114 $_{9}$	0.4731 $_{20}$	**218**
219	389.45 $_{39}$	362.6	1200.7	838.1	753.4	84.7	0.5524	2.105 $_{9}$	0.4751 $_{21}$	**219**
220	389.84 $_{39}$	363.0	1200.8	837.8	753.1	84.7	0.5529	2.096 $_{9}$	0.4772 $_{20}$	**220**
221	390.23 $_{39}$	363.5	1201.0	837.5	752.8	84.7	0.5533	2.087 $_{9}$	0.4792 $_{21}$	**221**
222	390.62 $_{39}$	363.9	1201.1	837.2	752.5	84.7	0.5538	2.078 $_{9}$	0.4813 $_{21}$	**222**
223	391.01 $_{39}$	364.3	1201.2	836.9	752.2	84.7	0.5543	2.069 $_{9}$	0.4834 $_{21}$	**223**
224	391.40 $_{39}$	364.7	1201.3	836.6	751.9	84.7	0.5548	2.060 $_{9}$	0.4855 $_{21}$	**224**
225	391.79 $_{38}$	365.1	1201.4	836.3	751.6	84.7	0.5553	2.051 $_{9}$	0.4876 $_{21}$	**225**
226	392.17 $_{38}$	365.5	1201.6	836.1	751.3	84.8	0.5557	2.042 $_{8}$	0.4896 $_{21}$	**226**
227	392.55 $_{38}$	365.9	1201.7	835.8	751.0	84.8	0.5562	2.034 $_{8}$	0.4917 $_{20}$	**227**
228	392.93 $_{38}$	366.3	1201.8	835.5	750.7	84.8	0.5567	2.026 $_{9}$	0.4939 $_{20}$	**228**
229	393.31 $_{38}$	366.7	1201.9	835.2	750.4	84.8	0.5571	2.017 $_{8}$	0.4959 $_{20}$	**229**
230	393.69 $_{38}$	367.1	1202.0	834.9	750.1	84.8	0.5576	2.009 $_{8}$	9.4979 $_{21}$	**230**
231	394.07 $_{3}$	367.5	1202.1	834.6	749.8	84.8	0.5581	2.001 $_{9}$	0.5000 $_{21}$	**231**
232	394.45 $_{3}$	367.9	1202.2	834.3	749.5	84.8	0.5585	1.992 $_{8}$	0.5021 $_{20}$	**232**
233	394.82 $_{37}$	368.3	1202.4	834.1	749.2	84.9	0.5590	1.984 $_{8}$	0.5041 $_{21}$	**233**

SATURATED STEAM—*Continued.*

Pressure, Pounds per Square Inch. p	Temperature, Degrees Fahr. t	Heat of the Liquid. q	Total Heat. λ	Heat of Vaporization. r	Heat equivalent of Internal Work. ρ	Heat equivalent of External Work. Apu	Entropy of the Liquid. $\int \frac{cdt}{T}$	Specific Volume. s	DENSITY. Weight, in Pounds, of one Cubic Foot. γ	Pressure, Pounds per Square Inch. p
234	395.19$_{37}$	368.6	1202.5	833.9	749.0	84.9	0.5594	1.976$_{8}$	0.5062$_{20}$	**234**
235	395.56$_{37}$	369.0	1202.6	833.6	748.7	84.9	0.5599	1.968$_{8}$	0.5082$_{21}$	**235**
236	395.93$_{37}$	369.4	1202.7	833.3	748.4	84.9	0.5603	1.960$_{8}$	0.5103$_{20}$	**236**
237	396.30$_{37}$	369.8	1202.8	833.0	748.1	84.9	0.5608	1.952$_{8}$	0.5123$_{21}$	**237**
238	396.67$_{37}$	370.2	1202.9	832.7	747.8	84.9	0.5612	1.944$_{8}$	0.5144$_{21}$	**238**
239	397.04$_{37}$	370.6	1203.0	832.4	747.5	84.9	0.5617	1.936$_{8}$	0.5165$_{21}$	**239**
240	397.41$_{36}$	371.0	1203.2	832.2	747.3	84.9	0.5621	1.928$_{7}$	0.5186$_{20}$	**240**
241	397.77$_{36}$	371.3	1203.3	832.0	747.0	85.0	0.5626	1.921$_{8}$	0.5206$_{20}$	**241**
242	398.13$_{36}$	371.7	1203.4	831.7	746.7	85.0	0.5630	1.913$_{7}$	0.5226$_{21}$	**242**
243	398.49$_{36}$	372.1	1203.5	831.4	746.4	85.0	0.5635	1.906$_{8}$	0.5247$_{21}$	**243**
244	398.85$_{36}$	372.5	1203.6	831.1	746.1	85.0	0.5639	1.898$_{7}$	0.5268$_{21}$	**244**
245	399.21$_{36}$	372.8	1203.7	830.9	745.9	85.0	0.5643	1.891$_{8}$	0.5289$_{22}$	**245**
246	399.57$_{36}$	373.2	1203.8	830.6	745.6	85.0	0.5648	1.883$_{8}$	0.5311$_{21}$	**246**
247	399.93$_{36}$	373.6	1203.9	830.3	745.3	85.0	0.5652	1.875$_{7}$	0.5332$_{21}$	**247**
248	400.29$_{35}$	374.0	1204.0	830.0	745.0	85.0	0.5656	1.868$_{7}$	0.5353$_{20}$	**248**
249	400.64$_{35}$	374.3	1204.1	829.8	744.8	85.0	0.5661	1.861$_{7}$	0.5373$_{20}$	**249**
250	400.99$_{35}$	374.7	1204.2	829.5	744.5	85.0	0.5665	1.854$_{7}$	0.5393$_{20}$	**250**
251	401.34$_{35}$	375.1	1204.4	829.3	744.2	85.1	0.5669	1.847$_{7}$	0.5413$_{20}$	**251**
252	401.69$_{35}$	375.4	1204.5	829.1	744.0	85.1	0.5673	1.840$_{7}$	0.5433$_{21}$	**252**
253	402.04$_{35}$	375.8	1204.6	828.8	743.7	85.1	0.5678	1.833$_{7}$	0.5454$_{21}$	**253**
254	402.39$_{35}$	376.2	1204.7	828.5	743.4	85.1	0.5682	1.826$_{7}$	0.5475$_{21}$	**254**
255	402.74$_{35}$	376.5	1204.8	828.3	743.2	85.1	0.5686	1.819$_{7}$	0.5496$_{21}$	**255**
256	403.09$_{35}$	376.9	1204.9	828.0	742.9	85.1	0.5690	1.812$_{7}$	0.5517$_{21}$	**256**
257	403.44$_{35}$	377.3	1205.0	827.7	742.6	85.1	0.5695	1.805$_{7}$	0.5538$_{21}$	**257**
258	403.79$_{34}$	377.6	1205.1	827.5	742.4	85.1	0.5699	1.798$_{6}$	0.5559$_{21}$	**258**
259	404.13$_{34}$	378.0	1205.2	827.2	742.1	85.1	0.5703	1.792$_{7}$	0.5580$_{21}$	**259**
260	404.47$_{34}$	378.4	1205.3	826.9	741.7	85.2	0.5707	1.785$_{6}$	0.5601$_{20}$	**260**
261	404.81$_{34}$	378.7	1205.4	826.7	741.5	85.2	0.5711	1.779$_{6}$	0.5621$_{21}$	**261**
262	405.15$_{34}$	379.1	1205.5	826.4	741.2	85.2	0.5715	1.773$_{7}$	0.5642$_{21}$	**262**
263	405.49$_{34}$	379.4	1205.6	826.2	741.0	85.2	0.5719	1.766$_{7}$	0.5663$_{21}$	**263**
264	405.83$_{34}$	379.8	1205.7	825.9	740.7	85.2	0.5724	1.759$_{6}$	0.5684$_{21}$	**264**
265	406.17$_{34}$	380.2	1205.8	825.6	740.4	85.2	0.5728	1.753$_{7}$	0.5705$_{21}$	**265**
266	406.51$_{33}$	380.5	1205.9	825.4	740.2	85.2	0.5732	1.746$_{6}$	0.5726$_{20}$	**266**
267	406.84$_{34}$	380.8	1206.0	825.2	740.0	85.2	0.5736	1.740$_{6}$	0.5746$_{21}$	**267**
268	407.18$_{34}$	381.2	1206.1	824.9	739.7	85.2	0.5740	1.734$_{6}$	0.5767$_{21}$	**268**
269	407.52$_{33}$	381.5	1206.2	824.7	739.5	85.2	0.5744	1.728$_{6}$	0.5788$_{21}$	**269**
270	407.85$_{33}$	381.9	1206.3	824.4	739.2	85.2	0.5748	1.722$_{6}$	0.5809$_{20}$	**270**
271	408.18$_{33}$	382.2	1206.4	824.2	739.0	85.2	0.5752	1.716$_{7}$	0.5829$_{21}$	**271**
272	408.51$_{33}$	382.6	1206.5	823.9	738.6	85.3	0.5756	1.709$_{6}$	0.5850$_{21}$	**272**
273	408.84$_{33}$	382.9	1206.6	823.7	738.4	85.3	0.5760	1.703$_{6}$	0.5871$_{21}$	**273**

Pressure, Pounds per Square Inch. p	Temperature, Degrees Fahr. t	Heat of the Liquid. q	Total Heat. λ	Heat of Vaporization. r	Heat equivalent of Internal Work. ρ	Heat equivalent of External Work. Apu	Entropy of the Liquid. $\int \frac{cdt}{T}$	Specific Volume. s	DENSITY. Weight, in Pounds, of one Cubic Foot. γ	Pressure, Pounds per Square Inch. p
274	409.17$_{33}$	383.3	1206.7	823.4	738.1	85.3	0.5764	1.697$_{6}$	0.5892$_{21}$	274
275	409.50$_{33}$	383.6	1206.8	823.2	737.9	85.3	0.5768	1.691$_{6}$	0.5913$_{21}$	275
276	409.83$_{33}$	384.0	1206.9	822.9	737.6	85.3	0.5772	1.685$_{6}$	0.5934$_{21}$	276
277	410.16$_{32}$	384.3	1207.0	822.7	737.4	85.3	0.5776	1.679$_{6}$	0.5955$_{21}$	277
278	410.48$_{32}$	384.6	1207.1	822.5	737.2	85.3	0.5779	1.673$_{5}$	0.5976$_{21}$	278
279	410.80$_{32}$	385.0	1207.2	822.2	736.9	85.3	0.5783	1.668$_{6}$	0.5997$_{23}$	279
280	411.12$_{33}$	385.3	1207.3	822.0	736.7	85.3	0.5787	1.662$_{6}$	0.602$_{2}$	280
281	411.44$_{32}$	385.6	1207.4	821.8	736.5	85.3	0.5791	1.656$_{6}$	0.604$_{2}$	281
282	411.76$_{32}$	386.0	1207.5	821.5	736.2	85.3	0.5795	1.650$_{5}$	0.606$_{2}$	282
283	412.08$_{32}$	386.3	1207.6	821.3	736.0	85.3	0.5799	1.645$_{6}$	0.608$_{2}$	283
284	412.40$_{32}$	386.6	1207.7	821.1	735.8	85.3	0.5803	1.639$_{5}$	0.610$_{2}$	284
285	412.72$_{32}$	387.0	1207.8	820.8	735.5	85.3	0.5806	1.634$_{6}$	0.612$_{2}$	285
286	413.04$_{32}$	387.3	1207.9	820.6	735.3	85.3	0.5810	1.628$_{5}$	0.614$_{2}$	286
287	413.36$_{32}$	387.7	1208.0	820.3	735.0	85.3	0.5814	1.623$_{6}$	0.616$_{2}$	287
288	413.68$_{32}$	388.0	1208.1	820.1	734.7	85.4	0.5818	1.617$_{5}$	0.618$_{2}$	288
289	414.00$_{32}$	388.3	1208.2	819.9	734.5	85.4	0.5822	1.612$_{5}$	0.620$_{2}$	289
290	414.32$_{31}$	388.6	1208.3	819.7	734.3	85.4	0.5826	1.607$_{6}$	0.622$_{3}$	290
291	414.63$_{31}$	389.0	1208.4	819.4	734.0	85.4	0.5829	1.601$_{5}$	0.625$_{2}$	291
292	414.94$_{31}$	389.3	1208.5	819.2	733.8	85.4	0.5833	1.596$_{5}$	0.627$_{2}$	292
293	415.25$_{31}$	389.6	1208.6	819.0	733.6	85.4	0.5837	1.591$_{6}$	0.629$_{2}$	293
294	415.56$_{31}$	390.0	1208.7	818.7	733.3	85.4	0.5840	1.585$_{5}$	0.631$_{2}$	294
295	415.87$_{31}$	390.3	1208.8	818.5	733.1	85.4	0.5844	1.580$_{5}$	0.633$_{2}$	295
296	416.18$_{31}$	390.6	1208.9	818.3	732.9	85.4	0.5848	1.575$_{5}$	0.635$_{2}$	296
297	416.49$_{31}$	390.9	1209.0	818.1	732.7	85.4	0.5851	1.570$_{6}$	0.637$_{2}$	297
298	416.80$_{31}$	391.3	1209.1	817.8	732.4	85.4	0.5855	1.564$_{5}$	0.639$_{2}$	298
299	417.11$_{31}$	391.6	1209.2	817.6	732.2	85.4	0.5859	1.559$_{5}$	0.641$_{3}$	299
300	417.42$_{30}$	391.9	1209.3	817.4	732.0	85.4	0.5863	1.554$_{5}$	0.644$_{2}$	300
301	417.72$_{30}$	392.2	1209.3	817.1	731.7	85.4	0.5866	1.549$_{5}$	0.646$_{2}$	301
302	418.02$_{30}$	392.5	1209.4	816.9	731.5	85.4	0.5870	1.544$_{5}$	0.648$_{2}$	302
303	418.32$_{30}$	392.8	1209.5	816.7	731.3	85.4	0.5873	1.539$_{5}$	0.650$_{2}$	303
304	418.62$_{30}$	393.2	1209.6	816.4	731.0	85.4	0.5877	1.534$_{5}$	0.652$_{2}$	304
305	418.92$_{30}$	393.5	1209.7	816.2	730.8	85.4	0.5880	1.529$_{5}$	0.654$_{2}$	305
306	419.22$_{30}$	393.8	1209.8	816.0	730.6	85.4	0.5884	1.524$_{4}$	0.656$_{2}$	306
307	419.52$_{30}$	394.1	1209.9	815.8	730.4	85.4	0.5888	1.520$_{5}$	0.658$_{2}$	307
308	419.82$_{30}$	394.4	1210.0	815.6	730.2	85.4	0.5891	1.515$_{5}$	0.660$_{2}$	308
309	420.12$_{30}$	394.7	1210.1	815.4	730.0	85.4	0.5895	1.510$_{5}$	0.662$_{2}$	309
310	420.42$_{30}$	395.0	1210.2	815.2	729.8	85.4	0.5898	1.505$_{5}$	0.664$_{2}$	310
311	420.72$_{30}$	395.4	1210.3	814.9	729.5	85.4	0.5902	1.500$_{4}$	0.666$_{2}$	311
312	421.02$_{30}$	395.7	1210.4	814.7	729.3	85.4	0.5905	1.496$_{5}$	0.668$_{3}$	312
313	421.32$_{30}$	396.0	1210.4	814.4	729.0	85.4	0.5909	1.491$_{5}$	0.671$_{2}$	313

SATURATED STEAM—*Continued.*

Pressure, Pounds per Square Inch. p	Temperature, Degrees Fahr. t	Heat of the Liquid. q	Total Heat. λ	Heat of Vaporization. r	Heat equivalent of Internal Work. ρ	Heat equivalent of External Work. Apu	Entropy of the Liquid. $\int \frac{cdt}{T}$	Specific Volume. s	Density. Weight, in Pounds, of one Cubic Foot. γ	Pressure, Pounds per Square Inch. p
314	421.62_{30}	396.3	1210.5	814.2	728.7	85.5	0.5913	1.486_{5}	0.673_{2}	**314**
315	421.92_{29}	396.6	1210.6	814.0	728.5	85.5	0.5916	1.481_{4}	0.675_{2}	**315**
316	422.21_{29}	396.9	1210.7	813.8	728.3	85.5	0.5919	1.477_{5}	0.677_{2}	**316**
317	422.50_{29}	397.2	1210.8	813.6	728.1	85.5	0.5923	1.472_{4}	0.679_{2}	**317**
318	422.79_{29}	397.5	1210.9	813.4	727.9	85.5	0.5926	1.468_{4}	0.681_{2}	**318**
319	423.08_{29}	397.8	1211.0	813.2	727.7	85.5	0.5930	1.464_{5}	0.683_{2}	**319**
320	423.37_{29}	398.1	1211.1	813.0	727.5	85.5	0.5933	1.459_{5}	0.685_{3}	**320**
321	423.66_{29}	398.4	1211.2	812.8	727.3	85.5	0.5937	1.454_{4}	0.688_{2}	**321**
322	423.95_{29}	398.7	1211.2	812.5	727.0	85.5	0.5940	1.450_{5}	0.690_{2}	**322**
323	424.24_{29}	399.0	1211.3	812.3	726.8	85.5	0.5944	1.445_{4}	0.692_{2}	**323**
324	424.53_{29}	399.3	1211.4	812.1	726.6	85.5	0.5947	1.441_{4}	0.694_{2}	**324**
325	424.82_{28}	399.6	1211.5	811.9	726.4	85.5	0.5950	1.437_{5}	0.696_{2}	**325**
326	425.10_{28}	399.9	1211.6	811.7	726.2	85.5	0.5954	1.432_{4}	0.698_{2}	**326**
327	425.38_{29}	400.2	1211.7	811.5	726.0	85.5	0.5957	1.428_{4}	0.700_{2}	**327**
328	425.67_{29}	400.5	1211.8	811.3	725.8	85.5	0.5960	1.424_{4}	0.702_{2}	**328**
329	425.96_{28}	400.8	1211.9	811.1	725.6	85.5	0.5964	1.420_{5}	0.704_{3}	**329**
330	426.24_{28}	401.1	1211.9	810.8	725.3	85.5	0.5967	1.415_{4}	0.707_{2}	**330**
331	426.52_{28}	401.4	1212.0	810.6	725.1	85.5	0.5970	1.411_{4}	0.709_{2}	**331**
332	426.80_{28}	401.7	1212.1	810.4	724.9	85.5	0.5974	1.407_{4}	0.711_{2}	**332**
333	427.08_{28}	402.0	1212.2	810.2	724.7	85.5	0.5977	1.403_{4}	0.713_{2}	**333**
334	427.36_{28}	402.3	1212.3	810.0	724.5	85.5	0.5980	1.399_{4}	0.715_{2}	**334**
335	427.64_{28}	402.6	1212.4	809.8	724.3	85.5	0.5984	1.395_{4}	0.717_{2}	**335**
336	427.92	402.9	1212.5	809.6	724.1	85.5	0.5987	1.391	0.719	**336**

TABLE III.

SATURATED STEAM.

FRENCH UNITS.

Temperature, Degrees Centigrade. t	Pressure, Millimeters of Mercury. p	Heat of the Liquid. q	Total Heat. λ	Heat of Vaporization. r	Heat equivalent of Internal Work. ρ	Heat equivalent of External Work. Apu	Entropy of the Liquid. $\int \frac{cdt}{T}$	Specific Volume. s	DENSITY. Weight, in Kilos, of one Cubic Meter. γ	Temperature, Degrees Centigrade. t
0	4.602_{339}	0.000	606.5	606.5	575.5	31.0	0.00000	211.5_{138}	0.004730_{327}	0
1	4.941_{362}	1.007	606.8	605.8	574.7	31.1	0.00367	197.7_{131}	0.005057_{360}	1
2	5.303_{386}	2.014	607.1	605.1	573.9	31.2	0.00733	184.6_{122}	0.005417_{383}	2
3	5.689_{411}	3.022	607.4	604.4	573.2	31.2	0.01098	172.4_{112}	0.005800_{403}	3
4	6.100_{436}	4.029	607.7	603.7	572.4	31.3	0.01461	161.2_{104}	0.006203_{427}	4
5	6.536_{465}	5.036	608.0	603.0	571.6	31.4	0.01823	150.8_{96}	0.006630_{450}	5
6	7.001_{493}	6.040	608.3	602.3	570.8	31.5	0.02183	141.2_{90}	0.007080_{481}	6
7	7.494_{525}	7.045	608.6	601.6	570.0	31.6	0.02542	132.2_{83}	0.007561_{508}	7
8	8.019_{557}	8.049	608.9	600.9	569.3	31.6	0.02899	123.9_{77}	0.008069_{529}	8
9	8.576_{591}	9.054	609.2	600.1	568.4	31.7	0.03255	116.2_{72}	0.008608_{569}	9
10	9.167_{628}	10.058	609.6	599.5	567.7	31.8	0.03609	109.0_{67}	0.009177_{602}	10
11	9.795_{665}	11.060	609.9	598.8	566.9	31.9	0.03962	102.3_{62}	0.009779_{631}	11
12	10.460_{704}	12.061	610.2	598.1	566.1	32.0	0.04313	96.09_{590}	0.01041_{67}	12
13	11.164_{747}	13.063	610.5	597.4	565.3	32.1	0.04663	90.19_{543}	0.01108_{71}	13
14	11.911_{791}	14.064	610.8	596.7	564.5	32.2	0.05012	84.76_{507}	0.01179_{76}	14
15	12.702_{837}	15.066	611.1	596.0	563.8	32.2	0.05359	79.69_{472}	0.01255_{79}	15
16	13.539_{884}	16.066	611.4	595.3	563.0	32.3	0.05705	74.97_{441}	0.01334_{83}	16
17	14.423_{937}	17.066	611.7	594.6	562.2	32.4	0.06050	70.56_{412}	0.01417_{88}	17
18	15.360_{989}	18.066	612.0	593.9	561.4	32.5	0.06330	66.44_{386}	0.01505_{93}	18
19	16.349_{1046}	19.066	612.3	593.2	560.6	32.6	0.06735	62.58_{360}	0.01598_{97}	19
20	17.395_{1103}	20.066	612.6	592.5	559.8	32.7	0.07076	58.98_{337}	0.01695_{103}	20
21	18.498_{1165}	21.064	612.9	591.8	559.0	32.8	0.07415	55.61_{315}	0.01798_{108}	21
22	19.663_{1229}	22.063	613.2	591.1	558.2	32.9	0.07754	52.46_{295}	0.01906_{114}	22
23	20.892_{1296}	23.061	613.5	590.4	557.5	32.9	0.08091	49.51_{277}	0.02020_{119}	23
24	22.188_{1366}	24.059	613.8	589.7	556.7	33.0	0.08427	46.74_{259}	0.02139_{126}	24
25	23.554_{1440}	25.058	614.1	589.0	555.9	33.1	0.08762	44.15_{243}	0.02265_{132}	25
26	24.994_{1516}	26.053	614.4	588.3	555.1	33.2	0.09094	41.72_{227}	0.02397_{138}	26
27	26.510_{1597}	27.048	614.7	587.7	554.4	33.3	0.09426	39.45_{214}	0.02535_{145}	27
28	28.107_{1679}	28.042	615.0	587.0	553.6	33.4	0.09756	37.31_{201}	0.02680_{153}	28
29	29.786_{1767}	29.037	615.3	586.3	552.8	33.5	0.10085	35.30_{188}	0.02833_{159}	29
30	31.553_{1858}	30.032	615.7	585.7	552.1	33.6	0.10413	33.42_{177}	0.02992_{168}	30

SATURATED STEAM—*Continued.*

Temperature, Degrees Centigrade. t	Pressure, Millimeters of Mercury. p	Heat of the Liquid. q	Total Heat. λ	Heat of Vaporization. r	Heat equivalent of Internal Work. ρ	Heat equivalent of External Work. Apu	Entropy of the Liquid. $\int \frac{cdt}{T}$	Specific Volume. s	Density. Weight, in Kilos, of one Cubic Meter. γ	Temperature, Degrees Centigrade. t
31	33.411_{1953}	31.027	616.0	585.0	551.3	33.7	0.10740	31.65_{167}	0.03160_{175}	31
32	35.364_{2052}	32.023	616.3	584.3	550.5	33.8	0.11067	29.98_{156}	0.03335_{184}	32
33	37.416_{2155}	33.018	616.6	583.6	549.7	33.9	0.11392	28.42_{148}	0.03519_{193}	33
34	39.571_{2262}	34.014	616.9	582.9	548.9	34.0	0.11716	26.94_{138}	0.03712_{201}	34
35	41.833_{2374}	35.009	617.2	582.2	548.2	34.0	0.12039	25.56_{131}	0.03913_{211}	35
36	44.207_{2490}	36.007	617.5	581.5	547.4	34.1	0.12362	24.25_{123}	0.04124_{220}	36
37	46.697_{2611}	37.005	617.8	580.8	546.6	34.2	0.12683	23.02_{116}	0.04344_{230}	37
38	49.308_{2742}	38.004	618.1	580.1	545.8	34.3	0.13004	21.86_{109}	0.04574_{241}	38
39	52.05_{286}	39.002	618.4	579.4	545.0	34.4	0.13324	20.77_{103}	0.04815_{251}	39
40	54.91_{301}	40.0	618.7	578.7	544.2	34.5	0.1364	19.74_{98}	0.05066_{263}	40
41	57.92_{314}	41.0	619.0	578.0	543.4	34.6	0.1396	18.76_{92}	0.05329_{275}	41
42	61.06_{329}	42.0	619.3	577.3	542.6	34.7	0.1428	17.84_{86}	0.05604_{285}	42
43	64.35_{345}	43.0	619.6	576.6	541.8	34.8	0.1459	16.98_{82}	0.05889_{298}	43
44	67.80_{360}	44.0	619.9	575.9	541.0	34.9	0.1491	16.16_{77}	0.06187_{310}	44
45	71.40_{376}	45.0	620.2	575.2	540.2	35.0	0.1522	15.39_{73}	0.06497_{325}	45
46	75.16_{394}	46.0	620.5	574.5	539.4	35.1	0.1554	14.66_{69}	0.06822_{338}	46
47	79.10_{411}	47.0	620.8	573.8	538.6	35.2	0.1585	13.97_{66}	0.07160_{352}	47
48	83.21_{430}	48.0	621.1	573.1	537.8	35.3	0.1617	13.31_{62}	0.07512_{366}	48
49	87.51_{447}	49.0	621.4	572.4	537.0	35.4	0.1648	12.69_{58}	0.07878_{381}	49
50	91.98_{467}	50.0	621.8	571.8	536.3	35.5	0.1679	12.11_{55}	0.08259_{394}	50
51	96.65_{489}	51.0	622.1	571.1	535.5	35.6	0.1710	11.56_{53}	0.08653_{416}	51
52	101.54_{510}	52.1	622.4	570.3	534.6	35.7	0.1741	11.03_{50}	0.09069_{428}	52
53	106.64_{531}	53.1	622.7	569.6	533.8	35.8	0.1772	10.53_{47}	0.09497_{443}	53
54	111.95_{554}	54.1	623.0	568.9	533.0	35.9	0.1803	10.06_{45}	0.09940_{470}	54
55	117.49_{576}	55.1	623.3	568.2	532.2	36.0	0.1833	9.610_{425}	0.1041_{48}	55
56	123.25_{601}	56.1	623.6	567.5	531.4	36.1	0.1864	9.185_{403}	0.1089_{50}	56
57	129.26_{625}	57.1	623.9	566.8	530.7	36.1	0.1895	8.782_{383}	0.1139_{52}	57
58	135.51_{651}	58.1	624.2	566.1	529.9	36.2	0.1925	8.399_{363}	0.1191_{54}	58
59	142.02_{678}	59.1	624.5	565.4	529.1	36.3	0.1956	8.036_{349}	0.1245_{56}	59
60	148.80_{705}	60.1	624.8	564.7	528.3	36.4	0.1986	7.687_{325}	0.1301_{57}	60
61	155.85_{733}	61.1	625.1	564.0	527.5	36.5	0.2016	7.362_{311}	0.1358_{60}	61
62	163.18_{762}	62.1	625.4	563.3	526.7	36.6	0.2046	7.051_{297}	0.1418_{63}	62
63	170.80_{792}	63.1	625.7	562.6	525.9	36.7	0.2076	6.754_{284}	0.1481_{65}	63
64	178.72_{823}	64.2	626.0	561.8	525.0	36.8	0.2106	6.470_{269}	0.1546_{67}	64
65	186.95_{855}	65.2	626.3	561.1	524.2	36.9	0.2136	6.201_{254}	0.1613_{69}	65
66	195.50_{888}	66.2	626.6	560.4	523.4	37.0	0.2166	5.947_{242}	0.1682_{71}	66
67	204.38_{922}	67.2	626.9	559.7	522.6	37.1	0.2196	5.705_{233}	0.1753_{74}	67
68	213.60_{957}	68.2	627.2	559.0	521.8	37.2	0.2225	5.472_{222}	0.1827_{78}	68
69	223.17_{992}	69.2	627.5	558.3	521.0	37.3	0.2254	5.250_{210}	0.1905_{80}	69
70	233.09_{1030}	70.2	627.9	557.7	520.3	37.4	0.2284	5.040_{201}	0.1985_{82}	70

SATURATED STEAM—*Continued.*

Temperature, Degrees Centigrade. t	Pressure, Millimeters of Mercury. p	Heat of the Liquid. q	Total Heat. λ	Heat of Vaporization. r	Heat equivalent of Internal Work. ρ	Heat equivalent of External Work. Apu	Entropy of the Liquid. $\int \frac{cdt}{T}$	Specific Volume. s	Density. Weight, in Kilos, of one Cubic Meter. γ	Temperature, Degrees Centigrade. t
71	243.39_{1068}	71.2	628.2	557.0	519.5	37.5	0.2313	4.839_{191}	0.2067_{84}	71
72	254.07_{1107}	72.2	628.5	556.3	518.7	37.6	0.2342	4.648_{183}	0.2151_{88}	72
73	265.14_{1148}	73.2	628.8	555.6	517.9	37.7	0.2371	4.465_{174}	0.2239_{91}	73
74	276.62_{1189}	74.2	629.1	554.9	517.1	37.8	0.2400	4.291_{167}	0.2330_{95}	74
75	288.51_{1232}	75.2	629.4	554.2	516.3	37.9	0.2429	4.124_{159}	0.2425_{97}	75
76	300.83_{1276}	76.2	629.7	553.5	515.5	38.0	0.2458	3.965_{152}	0.2522_{101}	76
77	313.59_{1321}	77.3	630.0	552.7	514.6	38.1	0.2487	3.813_{145}	0.2623_{103}	77
78	326.80_{1368}	78.3	630.3	552.0	513.8	38.2	0.2516	3.668_{139}	0.2726_{107}	78
79	340.48_{1415}	79.3	630.6	551.3	513.0	38.3	0.2544	3.529_{132}	0.2833_{111}	79
80	354.63_{1464}	80.3	630.9	550.6	512.3	38.3	0.2573	3.397_{127}	0.2944_{114}	80
81	369.27_{1514}	81.3	631.2	549.9	511.5	38.4	0.2601	3.270_{121}	0.3058_{118}	81
82	384.41_{1567}	82.3	631.5	549.2	510.7	38.5	0.2630	3.149_{116}	0.3176_{122}	82
83	400.08_{1619}	83.3	631.8	548.5	509.9	38.6	0.2658	3.033_{111}	0.3298_{125}	83
84	416.27_{1674}	84.3	632.1	547.8	509.1	38.7	0.2686	2.922_{107}	0.3423_{129}	84
85	433.01_{1730}	85.3	632.4	547.1	508.3	38.8	0.2714	2.815_{101}	0.3552_{133}	85
86	450.31_{1787}	86.3	632.7	546.4	507.5	38.9	0.2742	2.714_{98}	0.3685_{137}	86
87	468.18_{1846}	87.3	633.0	545.7	506.7	39.0	0.2770	2.616_{93}	0.3822_{143}	87
88	486.64_{1907}	88.3	633.3	545.0	505.9	39.1	0.2798	2.523_{90}	0.3965_{146}	88
89	505.71_{1969}	89.4	633.6	544.2	505.0	39.2	0.2826	2.433_{86}	0.4111_{149}	89
90	525.40_{2032}	90.4	634.0	543.6	504.3	39.3	0.2854	2.347_{82}	0.4260_{155}	90
91	545.72_{2098}	91.4	634.3	542.9	503.6	39.3	0.2881	2.265_{79}	0.4415_{160}	91
92	566.70_{2164}	92.4	634.6	542.2	502.8	39.4	0.2909	2.186_{76}	0.4575_{164}	92
93	588.34_{2233}	93.4	634.9	541.5	502.0	39.5	0.2937	2.110_{72}	0.4739_{169}	93
94	610.67_{2303}	94.4	635.2	540.8	501.2	39.6	0.2964	2.038_{70}	0.4908_{173}	94
95	633.70_{2375}	95.4	635.5	540.1	500.4	39.7	0.2991	1.968_{67}	0.5081_{180}	95
96	657.45_{2448}	96.4	635.8	539.4	499.6	39.8	0.3019	1.901_{65}	0.5261_{184}	96
97	681.93_{2524}	97.4	636.1	538.7	498.8	39.9	0.3046	1.836_{62}	0.5445_{191}	97
98	707.17_{2602}	98.4	636.4	538.0	498.1	39.9	0.3073	1.774_{59}	0.5636_{195}	98
99	733.19_{2681}	99.4	636.7	537.3	497.3	40.0	0.3100	1.715_{54}	0.5831_{191}	99
100	760.00_{275}	100.4	637.0	536.6	496.4	40.2	0.3127	1.661_{52}	0.6024_{195}	100
101	787.5_{283}	101.4	637.3	535.9	495.6	40.3	0.3154	1.609_{53}	0.6219_{208}	101
102	815.8_{292}	102.5	637.6	535.1	494.7	40.4	0.3181	1.556_{51}	0.6427_{218}	102
103	845.0_{301}	103.5	637.9	534.4	493.9	40.5	0.3208	1.505_{49}	0.6645_{223}	103
104	875.1_{309}	104.5	638.2	533.7	493.2	40.5	0.3235	1.456_{47}	0.6868_{229}	104
105	906.0_{319}	105.5	638.5	533.0	492.4	40.6	0.3261	1.409_{47}	0.7097_{236}	105
106	937.9_{328}	106.5	638.8	532.3	491.6	40.7	0.3288	1.365_{45}	0.7333_{243}	106
107	970.7_{337}	107.5	639.1	531.6	490.8	40.8	0.3314	1.320_{42}	0.7576_{249}	107
108	1004.4_{347}	108.5	639.4	530.9	490.1	40.8	0.3341	1.278_{40}	0.7825_{255}	108
109	1039.1_{356}	109.5	639.7	530.2	489.3	40.9	0.3367	1.248_{39}	0.8080_{260}	109
110	1074.7_{367}	110.5	640.1	529.6	488.6	41.0	0.3393	1.209_{37}	0.8340_{268}	110

SATURATED STEAM — *Continued.*

Temperature, Degrees Centigrade. t	Pressure, Millimeters of Mercury. p	Heat of the Liquid. q	Total Heat. λ	Heat of Vaporization. r	Heat equivalent of Internal Work. ρ	Heat equivalent of External Work. Apu	Entropy of the Liquid. $\int\frac{cdt}{T}$	Specific Volume. s	Density. Weight, in Kilos, of one Cubic Meter. γ	Temperature, Degrees Centigrade. t
111	1111.4_{377}	111.5	640.4	528.9	487.8	41.1	0.3420	1.162_{36}	0.8608_{275}	**111**
112	1149.1_{388}	112.5	640.7	528.2	487.0	41.2	0.3446	1.126_{35}	0.8883_{283}	**112**
113	1187.9_{398}	113.5	641.0	527.5	486.3	41.2	0.3471	1.091_{34}	0.9166_{290}	**113**
114	1227.7_{410}	114.6	641.3	526.7	485.4	41.3	0.3498	1.057_{32}	0.9456_{299}	**114**
115	1268.7_{420}	115.6	641.6	526.0	484.6	41.4	0.3524	1.025_{31}	0.9755_{305}	**115**
116	1310.7_{432}	116.6	641.9	525.3	483.8	41.5	0.3550	0.9942_{299}	1.006_{31}	**116**
117	1353.9_{444}	117.6	642.2	524.6	483.1	41.5	0.3576	0.9643_{289}	1.037_{32}	**117**
118	1398.3_{455}	118.6	642.5	523.9	482.3	41.6	0.3601	0.9354_{278}	1.069_{33}	**118**
119	1443.8_{467}	119.6	642.8	523.2	481.5	41.7	0.3627	0.9076_{268}	1.102_{33}	**119**
120	1490.5_{480}	120.6	643.1	522.5	480.7	41.8	0.3653	0.8808_{258}	1.135_{35}	**120**
121	1538.5_{492}	121.6	643.4	521.8	480.0	41.8	0.3678	0.8550_{250}	1.170_{35}	**121**
122	1587.7_{506}	122.6	643.7	521.1	479.2	41.9	0.3704	0.8300_{241}	1.205_{36}	**122**
123	1638.3_{518}	123.6	644.0	520.4	478.4	42.0	0.3729	0.8059_{233}	1.241_{37}	**123**
124	1690.1_{532}	124.6	644.3	519.7	477.6	42.1	0.3755	0.7826_{224}	1.278_{37}	**124**
125	1743.3_{545}	125.6	644.6	519.0	476.8	42.2	0.3780	0.7602_{216}	1.315_{39}	**125**
126	1797.8_{559}	126.6	644.9	518.3	476.1	42.2	0.3805	0.7386_{211}	1.354_{40}	**126**
127	1853.7_{573}	127.7	645.2	517.5	475.2	42.3	0.3830	0.7175_{203}	1.394_{40}	**127**
128	1911.0_{587}	128.7	645.5	516.8	474.4	42.4	0.3856	0.6973_{195}	1.434_{41}	**128**
129	1969.7_{601}	129.7	645.8	516.1	473.6	42.5	0.3881	0.6778_{187}	1.475_{42}	**129**
130	2029.8_{617}	130.7	646.2	515.5	473.0	42.5	0.3906	0.6591_{183}	1.517_{43}	**130**
131	2091.5_{633}	131.7	646.5	514.8	472.2	42.6	0.3931	0.6408_{177}	1.560_{45}	**131**
132	2154.8_{647}	132.7	646.8	514.1	471.4	42.7	0.3955	0.6231_{170}	1.605_{45}	**132**
133	2219.5_{663}	133.7	647.1	513.4	470.6	42.8	0.3980	0.6061_{165}	1.650_{46}	**133**
134	2285.8_{679}	134.7	647.4	512.7	469.8	42.9	0.4005	0.5896_{160}	1.696_{47}	**134**
135	2353.7_{695}	135.7	647.7	512.0	469.1	42.9	0.4030	0.5736_{153}	1.743_{48}	**135**
136	2423.2_{712}	136.7	648.0	511.3	468.3	43.0	0.4054	0.5583_{149}	1.791_{49}	**136**
137	2494.4_{728}	137.7	648.3	510.6	467.5	43.1	0.4079	0.5434_{145}	1.840_{51}	**137**
138	2567.2_{745}	138.7	648.6	509.9	466.7	43.2	0.4103	0.5289_{140}	1.891_{51}	**138**
139	2641.7_{762}	139.8	648.9	509.1	465.9	43.2	0.4128	0.5149_{136}	1.942_{53}	**139**
140	2717.9_{780}	140.8	649.2	508.4	465.1	43.3	0.4152	0.5013_{130}	1.995_{53}	**140**
141	2795.9_{798}	141.8	649.5	507.7	464.3	43.4	0.4177	0.4883_{127}	2.048_{55}	**141**
142	2875.7_{816}	142.8	649.8	507.0	463.5	43.5	0.4201	0.4756_{123}	2.103_{55}	**142**
143	2957.3_{835}	143.8	650.1	506.3	462.8	43.5	0.4225	0.4633_{119}	2.158_{57}	**143**
144	3040.8_{853}	144.8	650.4	505.6	462.0	43.6	0.4249	0.4514_{115}	2.215_{58}	**144**
145	3126.1_{872}	145.8	650.7	504.9	461.2	43.7	0.4273	0.4399_{112}	2.273_{59}	**145**
146	3213.3_{892}	146.8	651.0	504.2	460.4	43.8	0.4297	0.4287_{108}	2.232_{60}	**146**
147	3302.5_{911}	147.8	651.3	503.5	459.6	43.9	0.4321	0.4179_{105}	2.392_{62}	**147**
148	3393.6_{931}	148.8	651.6	502.8	458.9	43.9	0.4325	0.4074_{101}	2.454_{63}	**148**
149	3486.7_{952}	149.8	651.9	502.1	458.1	44.0	0.4369	0.3973_{98}	2.517_{64}	**149**
150	3581.9_{972}	150.8	652.3	501.5	457.4	44.1	0.4393	0.3875_{96}	2.581_{65}	**150**

SATURATED STEAM—*Continued.*

Temperature, Degrees Centigrade. t	Pressure in Millimeters of Mercury. p	Heat of the Liquid. q	Total Heat. λ	Heat of Vaporization. r	Heat equivalent of Internal Work. ρ	Heat equivalent of External Work. Apu	Entropy of the Liquid. $\int\frac{cdt}{T}$	Specific Volume. s	DENSITY. Weight, in Kilos, of a Cubic Meter. γ	Temperature, Degrees Centigrade. t
151	3679.1 993	151.8	652.6	500.8	456.6	2	0.4417	0.3779 93	2.646 67	**151**
152	3778.4 1014	152.9	652.9	500.0	455.8	44:2	0.4440	0.3686 90	2.713 68	**152**
153	3879.8 1035	153.9	653.2	499.3	455.0	44.3	0.4464	0.3596 87	2.781 69	**153**
154	3983.3 1057	154.9	653.5	498.6	454.2	4	0.4488	0.3509 85	2.850 70	**154**
155	4089,0 1079	155.9	653.8	497.9	453.4	44.5	0.4511	0.3424 82	2.920 72	**155**
156	4196.9 1102	156.9	654.1	497.2	452.7	44.5	0.4536	0.3342 80	2.992 74	**156**
157	4307.1 1124	158.0	654.4	496.4	451.8		0.4560	0.3262 78	3.066 75	**157**
158	4419.5 1148	159.0	654.7	495.7	450.0	44.6	0.4584	0.3184 76	3.141 76	**158**
159	4534.3 1171	160.1	655.0	494.9	449.2	44.7	0.4608	0.3108 73	3.217 78	**159**
160	4651.4 1195	161.1	655.3	494.2	449.4	44.8	0.4633	0.3035 71	3.295 79	**160**
161	4770.9 1218	162.2	655.6	493.4	448.5	9	0.4657	0.2964 69	3.374 80	**161**
162	4892.7 1243	163.2	655.9	492.7	447.7	44:0	0.4681	0.2895 67	3.454 82	**162**
163	5017. 127	164.2	656.2	492.0	447.0	45.0	0.4705	0.2828 66	3.536 84	**163**
164	5144. 129	165.3	656.5	491.2	446.1	.1	0.4729	0.2762 63	3.620 85	**164**
165	5273. 132	166.3	656.8	490.5	445.3	45.2	0.4752	0.2699 62	3.705 87	**165**
166	5405. 134	167.4	657.1	489.7	444.5	45.2	0.4776	0.2637 60	3.792 88	**166**
167	5539. 137	168.4	657.4	489.0	443.7	4 3	0.4800	0.2577 58	3.880 90	**167**
168	5676. 140	169.5	657.7	488.2	442.9	45.3	0.4824	0.2519 57	3.970 91	**168**
169	5816. 143	170.5	658.0	487.5	442.1	45.4	0.4847	0.2462 55	4.061 93	**169**
170	5959. 145	171.6	658.4	486.8	441.3	45.5	0.4871	0.2407 53	4.154 94	**170**
171	6104. 147	172.6	658.7	486.1	440.5	45.6	0.4895	0.2354 52	4.248 97	**171**
172	6251. 151	173.7	659.0	485.3	439.7	45.6	0.4918	0.2302 51	4.345 99	**172**
173	6402. 153	174.7	659.3	484.6	438.9	45.7	0.4941	0.2251 50	4.444 99	**173**
174	6555. 157	175.8	659.6	483.8	438.1	.7	0.4965	0.2201 48	4.543 101	**174**
175	6712. 159	176.8	659.9	483.1	437.3	4 .8	0.4988	0.2153 47	4.644 103	**175**
176	6871. 162	177.8	660.2	482.4	436.5	45.9	0.5011	0.2106 45	4.747 105	**176**
177	7033. 165	178.9	660.5	481.6	435.7	45.9	0.5035	0.2061 44	4.852 107	**177**
178	7198. 168	179.9	660.8	480.9	434.9	46.0	0.5058	0.2017 44	4.959 109	**178**
179	7366. 171	181.0	661.1	480.1	434.0	46.1	0.5081	0.1973 42	5.068 110	**179**
180	7537. 175	182.0	661.4	479.4	433.3	46.1	0.5104	0.1931 41	5.178 113	**180**
181	7712. 177	183.1	661.7	478.6	432.4	46.2	0.5127	0.1890 40	5.291 114	**181**
182	7889. 181	184.1	662.0	477.9	431.7	46.2	0.5150	0.1850 39	5.405 117	**182**
183	8070. 183	185.2	662.3	477.1	430.8	46.3	0.5173	0.1811 38	5.522 118	**183**
184	8253. 187	186.2	662.6	476.4	430.1	46.3	0.5196	0.1773 37	5.640 120	**184**
185	8440. 191	187.3	662.9	475.6	429.2	46.4	0.5219	0.1736 36	5.760 122	**185**
186	8631. 193	188.3	663.2	474.9	428.5	46.4	0.5242	0.1700 36	5.882 125	**186**
187	8824. 197	189.4	663.5	474.1	427.6	46.5	0.5264	0.1664 34	6.007 127	**187**
188	9021. 201	190.4	663.8	473.4	426.9	46.5	0.5287	0.1630 33	6.134 128	**188**
189	9222. 204	191.4	664.1	472.7	426.1	46.6	0.5310	0.1597 33	6.262 130	**189**
190	9426. 207	192.5	664.5	472.0	425.3	46.7	0.5332	0.1564 32	6.392 133	**190**

SATURATED STEAM—*Concluded.*

Temperature, Degrees Centigrade. t	Pressure, Millimeters of Mercury. p	Heat of the Liquid. q	Total Heat. λ	Heat of Vaporization. r	Heat equivalent of Internal Work. ρ	Heat equivalent of External Work. Apu	Entropy of the Liquid. $\int\frac{cdt}{T}$	Specific Volume. s	DENSITY. Weight, in Kilos, of one Cubic Meter. γ	Temperature, Degrees Centigrade. t
191	9633.$_{211}$	193.5	664.8	471.3	424.6	46.7	0.5355	0.1532$_{31}$	6.525$_{136}$	**191**
192	9844.$_{214}$	194.6	665.1	470.5	423.7	46.8	0.5377	0.1501$_{30}$	6.661$_{137}$	**192**
193	10058.$_{218}$	195.6	665.4	469.8	423.0	46.8	0.5400	0.1471$_{30}$	6.798$_{140}$	**193**
194	10276.$_{222}$	196.7	665.7	469.0	422.2	46.8	0.5422	0.1441$_{29}$	6.938$_{142}$	**194**
195	10498.$_{226}$	197.7	666.0	468.3	421.4	46.9	0.5444	0.1412$_{28}$	7.080$_{145}$	**195**
196	10724.$_{229}$	198.8	666.3	467.5	420.6	46.9	0.5467	0.1384$_{27}$	7.225$_{147}$	**196**
197	10953.$_{233}$	199.8	666.6	466.8	419.8	47.0	0.5489	0.1357$_{27}$	7.372$_{149}$	**197**
198	11186.$_{238}$	200.9	666.9	466.0	419.0	47.0	0 5511	0.1330$_{27}$	7.521$_{151}$	**198**
199	11424.$_{240}$	201.9	667.2	465.3	418.2	47.1	0.5533	0.1303$_{26}$	7.672$_{155}$	**199**
200	11664.$_{245}$	203.0	667.5	464.5	417.4	47.1	0.5555	0.1277$_{25}$	7.827$_{157}$	**200**
201	11909.$_{249}$	204.0	667.8	463.8	416.7	47.1	0.5577	0.1252$_{24}$	7.984$_{159}$	**201**
202	12158.$_{253}$	205.0	668.1	463.1	415.9	47.2	0.5599	0.1228$_{24}$	8.143$_{162}$	**202**
203	12411.$_{257}$	206.1	668.4	462.3	415.1	47.2	0.5621	0.1204$_{23}$	8.305$_{165}$	**203**
204	12668.$_{262}$	207.1	668.7	461.6	414.4	47.2	0.5643	0.1181$_{23}$	8.470$_{169}$	**204**
205	12930.$_{265}$	208.2	669.0	460.8	413.5	47.3	0.5665	0.1158$_{23}$	8.639$_{171}$	**205**
206	13195.$_{270}$	209.2	669.3	460.1	412.8	47.3	0.5687	0.1135$_{22}$	8.810$_{174}$	**206**
207	13465.$_{274}$	210.3	669.6	459.3	412.0	47.3	0.5709	0.1113$_{21}$	8.984$_{176}$	**207**
208	13739.$_{279}$	211.3	669.9	458.6	411.3	47.3	0.5731	0.1092$_{21}$	9.160$_{178}$	**208**
209	14018.$_{283}$	212.4	670.2	457.8	410.4	47.4	0.5752	0.1071$_{21}$	9.338$_{181}$	**209**
210	14301.$_{287}$	213.4	670.6	457.2	409.8	47.4	0.5774	0.1050$_{20}$	9.519$_{185}$	**210**
211	14588.$_{292}$	214.5	670.9	456.4	409.0	47.4	0.5795	0.1030$_{19}$	9.704$_{190}$	**211**
212	14880.$_{297}$	215.5	671.2	455.7	408.3	47.4	0.5817	0.1011$_{19}$	9.894$_{186}$	**212**
213	15177.$_{301}$	216.5	671.5	455.0	407.6	47.4	0.5839	0.0992$_{19}$	10.08$_{20}$	**213**
214	15478.$_{307}$	217.6	671.8	454.2	406.7	47.5	0.5860	0.0973$_{19}$	10.28$_{20}$	**214**
215	15785.$_{311}$	218.6	672.1	453.5	406.0	47.5	0.5881	0.0954$_{18}$	10.48$_{20}$	**215**
216	16096.$_{315}$	219.7	672.4	452.7	405.2	47.5	0.5903	0.0936$_{18}$	10.68$_{21}$	**216**
217	16411.$_{321}$	220.7	672.7	452.0	404.5	47.5	0.5924	0.0918$_{17}$	10.89$_{21}$	**217**
218	16732.$_{326}$	221.8	673.0	451.2	403.7	47.5	0.5945	0.0901$_{17}$	11.10$_{21}$	**218**
219	17058.$_{331}$	222.8	673.3	450.5	403.0	47.5	0.5967	0.0884$_{16}$	11.31$_{22}$	**219**
220	17389.	223.9	673.6	449.7	402.2	47.5	0.5988	0.0868	11.53	**220**

TABLE IV.

SATURATED VAPOR OF ETHER.

FRENCH UNITS.

Temperature, Degrees Centigrade. t	Pressure, Millimeters of Mercury. p	Heat of the Liquid. q	Total Heat. λ	Heat of Vaporization. r	Heat equivalent of Internal Work. ρ	Heat equivalent of External Work. Apu	Entropy of the Liquid. $\int \frac{cdt}{T}$	Specific Volume. s	DENSITY. Weight, in Kilos, of one Cubic Meter. γ	Temperature, Degrees Centigrade. t
0	184.39	0.00	94.00	94.00	86.45	7.55	0.0000	1.278	0.782	0
10	286.83	5.32	98.44	93.12	85.37	7.75	0.01909	0.8440	1.185	10
20	432.78	10.70	102.78	92.08	84.13	7.95	0.03772	0.5741	1.742	20
30	634.80	16.14	107.00	90.86	82.72	8.14	0.05593	0.4013	2.492	30
40	907.04	21.63	111.11	89.48	81.15	8.33	0.07374	0.2877	3.746	40
50	1264.8	27.19	115.11	87.92	79.41	8.51	0.09117	0.2108	4.744	50
60	1725.0	32.80	119.00	86.20	77.53	8.67	0.1083	0.1580	6.329	60
70	2304.9	38.48	122.78	84.30	75.49	8.81	0.1250	0.1203	8.313	70
80	3022.8	44.21	126.44	82.23	73.32	8.91	0.1415	0.0932	10.73	80
90	3898.3	50.00	130.00	80.00	71.03	8.97	0.1576	0.0731	13.68	90
100	4953.3	55.86	133.44	77.58	68.62	8.96	0.1735	0.0577	17.33	100
110	6214.6	61.77	136.78	75.01	66.13	8.88	0.1891	0.0459	21.79	110
120	7719.2	67.74	140.00	72.26	63.57	8.69	0.2045	0.0364	27.47	120

TABLE V.

SATURATED VAPOR OF ALCOHOL.

FRENCH UNITS.

Temperature, Degrees Centigrade. t	Pressure per Millimeters of Mercury. p	Heat of the Liquid. q	Total Heat. λ	Heat of Vaporization. r	Heat equivalent of Internal Work. ρ	Heat equivalent of External Work. Apu	Entropy of the Liquid. $\int \frac{cdt}{T}$	Specific Volume. s	DENSITY. Weight, in Kilos, of one Cubic Meter. γ	Temperature, Degrees Centigrade. t
0	12.70	0.00	236.5	236.50	223.38	13.12	0.0000	32.21	0.03105	0
10	24.23	5.59	244.4	238.81	225.29	13.52	0.01996	17.39	0.05750	10
20	44.46	11.42	252.0	240.58	226.56	14.02	0.04003	9.847	0.1016	20
30	78.52	17.49	258.0	240.51	226.03	14.48	0.06029	5.753	0.1738	30
40	133.69	23.71	262.0	238.29	223.44	14.85	0.08073	3.465	0.2886	40
50	219.90	30.21	264.0	233.79	218.59	15.10	0.1014	2.143	0.4666	50
60	350.21	37.37	265.0	227.63	212.38	15.25	0.1223	1.359	0.7358	60
70	541.15	44.58	265.2	220.62	205.28	15.34	0.1435	0.8855	1.129	70
80	812.91	52.11	265.2	213.09	197.69	15.40	0.1650	0.5921	1.689	80
90	1189.3	59.97	266.0	206.03	190.54	15.49	0.1868	0.4073	2.455	90
100	1697.6	68.18	267.3	199.12	183.54	15.58	0.2090	0.2874	3.479	100
110	2367.6	76.74	269.6	192.86	177.15	15.71	0.2315	0.2083	4.801	110
120	3231.7	85.67	272.5	186.83	170.97	15.86	0.2544	0.1544	6.477	120
130	4323.0	94.98	276.0	181.02	164.99	16.03	0.2776	0.1170	8.547	130
140	5674.6	104.70	280.5	175.80	159.55	16.25	0.3013	0.0905	11.05	140
150	7318.4	114.82	285.3	170.48	154.03	16.45	0.3254	0.0714	14.01	150

TABLE VI.

SATURATED VAPOR OF CHLOROFORM.

FRENCH UNITS.

Temperature, Degrees Centigrade. t	Pressure, Millimeters of Mercury. p	Heat of the Liquid. q	Total Heat. λ	Heat of Vaporization. r	Heat equivalent of Internal Work. ρ	Heat equivalent of External Work. Apu	Entropy of the Liquid. $\int \frac{cdt}{T}$	Specific Volume. s	DENSITY. Weight, in Kilos, of one Cubic Meter. γ	Temperature, Degrees Centigrade. t
0	59.72	0.00	67.00	67.00	62.45	4.55	0.00000	2.377	0.4207	0
10	100.47	2.33	68.38	66.04	61.29	4.75	0.00836	1.475	0.6780	10
20	160.47	4.67	69.75	65.08	60.14	4.94	0.01646	0.9601	1.042	20
30	247.51	7.02	71.12	64.10	59.00	5.10	0.02432	0.6437	1.554	30
40	369.26	9.37	72.50	63.13	57.87	5.26	0.03196	0.4449	2.248	40
50	535.05	11.74	73.87	62.13	56.73	5.40	0.03940	0.3155	3.170	50
60	755.44	14.12	75.25	61.13	55.60	5.53	0.04664	0.2291	4.356	60
70	1042.1	16.51	76.62	60.11	54.45	5.66	0.05369	0.1700	5.88	70
80	1407.6	18.91	78.00	59.09	53.31	5.78	0.06057	0.1286	7.78	80
90	1865.2	21.32	79.37	58.05	52.16	5.89	0.06729	0.0991	10.09	90
100	2428.5	23.74	80.75	57.01	51.01	6.00	0.07386	0.0777	12.87	100
110	3111.0	26.17	82.12	55.95	49.84	6.11	0.08027	0.0618	16.18	110
120	3925.7	28.61	83.50	54.89	48.67	6.22	0.08655	0.0500	20.00	120
130	4885.1	31.06	84.87	53.81	47.48	6.33	0.09270	0.0410	24.39	130
140	6000.2	33.52	86.25	52.73	46.30	6.43	0.09872	0.0340	29.4	140
150	7280.6	35.99	87.62	51.63	45.10	6.53	0.10462	0.0286	35.0	150
160	8734.2	38.47	89.00	50.53	43.90	6.63	0.11041	0.0243	41.2	160

TABLE VII.

SATURATED VAPOR OF CARBON BISULPHIDE.

FRENCH UNITS.

Temperature, Degrees Centigrade. t	Pressure, Millimeters of Mercury. p	Heat of the Liquid. q	Total Heat. λ	Heat of Vaporization. r	Heat equivalent of Internal Work. ρ	Heat equivalent of External Work. Apu	Entropy of the Liquid. $\int\frac{cdt}{T}$	Specific Volume. s	DENSITY. Weight, in Kilos, of one Cubic Meter. γ	Temperature, Degrees Centigrade. t
0	127.91	0.00	90.00	90.00	82.76	7.24	0.00000	1.766	0.5662	0
10	198.46	2.36	91.42	89.06	81.58	7.48	0.00847	1.177	0.8496	10
20	298.03	4.74	92.76	88.02	80.31	7.71	0.01670	0.8071	1.239	20
30	434.62	7.13	94.01	86.88	78.97	7.91	0.02472	0.5684	1.759	30
40	617.53	9.54	95.18	85.64	77.54	8.10	0.03252	0.4098	2.440	40
50	857.07	11.96	96.27	84.31	76.04	8.27	0.04013	0.3017	3.315	50
60	1164.5	14.41	97.28	82.87	74.45	8.42	0.04756	0.2264	4.417	60
70	1552.1	16.86	98.20	81.34	72.78	8.56	0.05482	0.1726	5.794	70
80	2032.5	19.34	99.04	79.70	71.03	8.67	0.06192	0.1338	7.473	80
90	2619.1	21.83	99.80	77.97	69.20	8.77	0.06886	0.1052	9.51	90
100	3325.2	24.34	100.48	76.14	67.29	8.85	0.07566	0.0837	11.95	100
110	4164.1	26.86	101.07	74.21	65.31	8.90	0.08233	0.0674	14.84	110
120	5148.8	29.40	101.58	72.18	63.24	8.94	0.08886	0.0549	18.21	120
130	6291.6	31.96	102.01	70.05	61.09	8.96	0.09527	0.0452	22.12	130
140	7604.0	34.53	102.36	67.83	58.88	8.95	0.10157	0.0375	26.7	140
150	9095.9	37.12	102.62	65.50	56.58	8.92	0.10775	0.0314	31.8	150

TABLE VIII.

SATURATED VAPOR OF CARBON TETRACHLORIDE.

FRENCH UNITS.

Temperature, Degrees Centigrade. t	Pressure, Millimeters of Mercury. p	Heat of the Liquid. q	Total Heat. λ	Heat of Vaporization. r	Heat equivalent of Internal Work. ρ	Heat equivalent of External Work. Apu	Entropy of the Liquid. $\int\frac{cdt}{T}$	Specific Volume. s	DENSITY. Weight, in Kilos, of one Cubic Meter. γ	Temperature, Degrees Centigrade. t
0	32.95	0.00	52.00	52.00	48.54	3.46	0.00000	3.272	0.3056	**0**
10	55.97	1.99	53.44	51.45	47.85	3.60	0.00714	2.005	0.4987	**10**
20	90.99	3.99	54.86	50.87	47.13	3.74	0.01409	1.283	0.7794	**20**
30	142.27	6.02	56.23	50.21	46.33	3.88	0.02087	0.8510	1.175	**30**
40	214.81	8.06	57.58	49.52	45.51	4.01	0.02749	0.5831	1.715	**40**
50	314.38	10.12	58.88	48.76	44.62	4.14	0.03396	0.4109	2.434	**50**
60	447.43	12.20	60.16	47.96	43.69	4.25	0.04028	0.2969	3.368	**60**
70	621.15	14.30	61.40	47.10	42.75	4.35	0.04648	0.2192	4.562	**70**
80	843.29	16.42	62.60	46.18	41.74	4.44	0.04255	0.1650	6.061	**80**
90	1122.3	18.55	63.77	45.22	40.50	4.72	0.05849	0.1263	7.92	**90**
100	1467.1	20.70	64.90	44.20	39.62	4.58	0.06433	0.0980	10.20	**100**
110	1887.4	22.87	66.01	43.14	38.52	4.62	0.07006	0.0770	12.99	**110**
120	2393.7	25.06	67.07	42.01	37.36	4.65	0.07569	0.0611	16.37	**120**
130	2996.9	27.27	68.10	40.83	36.18	4.65	0.08122	0.0490	20.41	**130**
140	3709.0	29.49	69.10	39.61	34.95	4.63	0.08666	0.0395	25.3	**140**
150	4543.1	31.73	70.07	38.34	33.75	4.59	0.09201	0.0321	31.2	**150**
160	5513.1	34.00	71.00	37.00	32.47	4.53	0.09729	0.0262	38.2	**160**

TABLE IX.

SATURATED VAPOR OF ACETON.

FRENCH UNITS.

Temperature, Degrees Centigrade. t	Pressure, Millimeters of Mercury. p	Heat of the Liquid. q	Total Heat. λ	Heat of Vaporization. r	Heat equivalent of Internal Work. ρ	Heat equivalent of External Work. Apu	Entropy of the Liquid. $\int \frac{cdt}{T}$	Specific Volume. s	DENSITY. Weight, in Kilos, of one Cubic Meter. γ	Temperature, Degrees Centigrade. t
0	63.33	0.00	140.50	140.50	131.82	8.68	0.00000	4.275	0.2339	0
10	110.32	5.10	144.11	139.01	129.51	9.50	0.01832	2.686	0.3723	10
20	180.08	10.29	147.62	137.33	127.16	10.17	0.03627	1.758	0.5688	20
30	280.05	15.55	151.03	135.48	124.83	10.65	0.05389	1.187	0.8425	30
40	419.35	20.89	154.33	133.44	121.39	11.05	0.07119	0.8227	1.215	40
50	608.81	26.31	157.53	131.22	119.86	11.36	0.08820	0.5830	1.715	50
60	860.96	31.81	160.63	128.82	117.22	11.60	0.1049	0.4215	2.372	60
70	1189.9	37.39	163.62	126.23	114.43	11.80	0.1214	0.3106	3.220	70
80	1611.1	43.05	166.51	123.46	111.49	11.97	0.1376	0.2328	4.296	80
90	2140.8	48.79	169.30	120.51	108.41	12.10	0.1536	0.1773	5.640	90
100	2796.2	54.61	171.98	117.37	105.17	12.20	0.1694	0.1372	7.289	100
110	3594.3	60.50	174.56	114.06	101.78	12.28	0.1850	0.1076	9.294	110
120	4552.0	66.48	177.04	110.56	98.23	12.33	0.2004	0.0856	11.68	120
130	5684.9	72.54	179.42	106.88	94.53	12.35	0.2156	0.0689	14.51	130
140	7007.6	78.67	181.69	103.02	90.67	12.35	0.2306	0.0561	17.83	140

TABLE X.

SATURATED VAPOR OF SULPHUR DIOXIDE.

FRENCH UNITS.

Temperature, Degrees Centigrade. t	Pressure, Kilograms per Square Meter. p	Heat of the Liquid. q	Total Heat. λ	Heat of Vaporization. r	Heat equivalent of Internal Work. ρ	Heat equivalent of External Work. Apu	Specific Volume. s	DENSITY. Weight, in Kilos, of one Cubic Meter. γ	Temperature, Degrees Centigrade. t
−30	3.91	−10.9	87.5	98.3	90.87	7.6	0.823	1.22	−30
−25	5.08	− 9.1	88.1	97.2	89.51	7.7	0.642	1.56	−25
−20	6.51	− 7.3	88.8	96.1	88.35	7.8	0.508	1.97	−20
−15	8.27	− 5.4	89.5	94.9	87.0	7.9	0.407	2.46	−15
−10	10.37	− 3.6	90.1	93.7	85.7	8.0	0.329	3.04	−10
− 5	12.87	− 1.8	90.8	92.6	84.4	8.1	0.269	3.72	− 5
0	15.84	0.0	91.4	91.4	83.2	8.2	0.221	4.52	0
5	19.32	1.8	92.0	90.2	81.9	8.3	0.184	5.45	5
10	23.38	3.6	92.4	89.0	80.6	8.4	0.154	6.51	10
15	28.07	5.5	93.3	87.8	79.3	8.5	0.129	7.74	15
20	33.47	7.3	93.9	86.6	78.0	8.6	0.110	9.13	20
25	39.65	9.1	94.5	85.4	76.7	8.7	0.0933	10.7	25
30	46.67	10.9	95.1	84.2	75.4	8.7	0.0800	12.5	30
35	54.59	12.8	95.5	82.9	74.1	8.8	0.0689	14.5	35
40	63.50	14.6	96.3	81.7	72.8	8.8	0.0597	16.8	40

TABLE XI.

SATURATED VAPOR OF AMMONIA.

FRENCH UNITS.

Temperature, Degrees Centigrade. t	Pressure, Kilograms per Square Meter. p	Heat of the Liquid. q	Total Heat. λ	Heat of Vaporization. r	Heat equivalent of Internal Work. ρ	Heat equivalent of External Work. Apu	Specific Volume. s	DENSITY. Weight, in Kilos, of one Cubic Meter. γ	Temperature, Degrees Centigrade. t
−40	7.19	−37.3	298.0	335.2	308.7	26.5	1.56	0.639	−40
−35	9.30	−33.0	300.0	332.9	305.9	27.0	1.23	0.812	−35
−30	11.92	−28.5	302.0	330.5	303.0	27.4	0.978	1.02	−30
−25	15.12	−23.0	303.9	327.0	300.0	27.9	0.784	1.28	−25
−20	19.00	−19.4	305.9	325.3	296.9	28.9	0.634	1.58	−20
−15	23.67	−14.7	307.9	322.5	293.7	28.8	0.518	1.93	−15
−10	29.23	− 9.9	309.8	319.7	290.4	29.3	0.426	2.35	−10
− 5	35.80	− 5.0	311.7	316.7	287.0	29.7	0.352	2.84	− 5
0	43.480	0.0	313.6	313.6	283.5	30.2	0.295	3.39	0
5	52.41	5.1	315.5	310.5	279.9	30.6	0.249	4.02	5
10	62.71	10.2	317.4	307.2	276.1	31.0	0.211	4.74	10
15	74.50	15.5	319.3	303.8	272.3	31.5	0.180	5.54	15
20	87.93	20.8	321.1	300.3	268.4	31.9	0.155	6.45	20
25	103.07	26.3	323.0	296.7	264.4	32.3	0.134	7.45	25
30	120.08	31.8	324.8	293.0	269.2	32.8	0.117	8.55	30
35	139.05	37.4	326.6	289.2	256.0	33.2	0.102	9.77	35
40	160.11	43.2	328.4	285.2	251.6	33.6	0.090	11.07	40

TABLE XII.

SPECIFIC GRAVITY AND SPECIFIC VOLUME OF LIQUIDS.

Name of Liquid.	Specific Gravity, compared with Water at 4° C.	Specific Volume. Cubic Meters per Kilo.
Alcohol, C_2H_6O	0.80625 [Mendelejeff, 1869]	0.001240
Ether, $C_4H_{10}O$	0.736 [Kopp, 1860]	0.001358
Chloroform	1.527 [Thorpe, 1880]	0.000655
Carbon bisulphide, CS_2	1.2922 [Thorpe, 1880]	0.000774
Carbon tetrachloride, CCl_4	1.6320 [Thorpe, 1880]	0.000613
Aceton, C_3H_6O	0.81 [Zander, 1882]	0.00123
Sulphur Dioxide SO_2	1.4336 [Andréeff, 1859]	0.0006981
Ammonia NH_3	0.6364 [Andréeff, 1859]	0.001571

TABLE XIII.

VOLUME OF WATER.

Vol. at 4° C=1.

[Rossetti, 1871] and [Hirn, 1867.]

Temperature.	Volume.	Temperature.	Volume.	Temperature.	Volume.	Temperature.	Volume.
10	1.000253	**60**	1.01691	**110**	1.0512	**160**	1.1018
20	1.001744	**70**	1.02256	**120**	1.0599	**170**	1.1139
30	1.00425	**80**	1.02887	**130**	1.0694	**180**	1.1268
40	1.00770	**90**	1.03567	**140**	1.0795	**190**	1.1403
50	1.01195	**100**	1.04312	**150**	1.0903	**200**	1.1544

CPSIA information can be obtained
at www.ICGtesting.com
Printed in the USA
LVOW04s1039020916
502977LV00020B/389/P

9 781332 759309